# THE DEVELOPMENT AND ANALYSIS
# OF
# NEW CHEMICAL PLANTS
# AND
# PROCESSES

by

# B. N. Nnolim

First published November 2001
By
CECTA (NIG) LIMITED
159 Chime Avenue,
New Haven, Enugu, Nigeria

ISBN 978-35581-6-1

Paperback Edition

ISBN 978-1-906914-48-6

Other Books by Ben Nnolim Books

|  | Paperback | eBook |
|---|---|---|
| Applied Heat Transfer (With Worked Examples) Volume 1: Conduction of Heat in Solids | 978-1-906914-21-9 | 978-1-906914-26-4 |
| Applied Heat Transfer Volume Two (With Worked Examples) Heat Convection in Fluids | 978-1-906914-22-6 | 978-1-906914-25-7 |
| Fundamentals of Mass Transfer | 978-1-906914-01-1 | 978-1-906914-27-1 |
| Worked Examples in Mass Transfer | 978-1-906914-46-2 | 978-1-906914-47-9 |
| Technological Emancipation of Nigeria- The Role of Chemical Engineering (A Collection of Lectures) | 978-1-906914-18-9 | 978-1-906914-34-9 |

Ben Nnolim Books
7 Sandway Path,
St. Mary Cray,
Orpington, Kent
BR5 3TL
Email: bnbs@ymail.com

# PREFACE

The purpose of this book is to make available to engineers and technologists in developing countries, who wish to develop and/or design manufacturing plants or processes for their indigenous products or plants for Western European lifestyle products, a straightforward, non-intimidating version of modern chemical engineering methods and analyses for such purposes.

This book owes a great deal to such standard text in chemical engineering as Perry's *Chemical Engineers' Handbook*, 6th. edition, edited by Don Green; *Chemical Plant Design and Economics* by Peters and Timmerhaus; *Chemical Engineering and Computer Calc*ulations by Myers and Syder; *Chemical Engineering* by Coulson and Richardson, Volume 6, edited by Sinnott; *Chemical Plant Design* by Vilbrandt and Dryden; *Chemical Process Principles, Volume 1*, by Hougen, Watson and Ragatz; *Material and Energy Balances* by Schmidt and List; various volumes of the *Chemical Engineering* magazine published by the McGraw-Hill Book Company, USA; *ChemTech*, edited by Peter Luberoff for the American Chemical Society and a host of others listed under references at the end of each chapter.

Grateful acknowledgements, for the physical and thermodynamic property data in the Appendices, are due to the trade publication *Physical and Thermodynamic Properties of Elements and Compounds* by the Chemetron Corporation, Catalysts Division and to *www.engineeringtoolbox.com*.

In this age of passing fads and fancies, one of which is the fashionable doubting of the existence of God, and confusing Christianity with its minority of less than genuine practitioners, I am happy and proud to express my gratitude to Our Blessed Virgin Mary, the Immaculate Conception, Help of Christians, Theotokos and the Queen of Heaven, who has always been a real mother to me and to God whose greatness, kindness and love overwhelm. I pray that this book and all my life are to His purpose.

B. N. Nnolim
October 3, 2011

# TABLE OF CONTENTS

PREFACE ...................................................................................II

CHAPTER ONE:  FROM IDEA TO PROCESS TO PRODUCT ......................1

1.0: FROM IDEA TO PRODUCT ..............................................................1
1.1: COMPANIES ...........................................................................1
1.2: THE CHEMICAL MANUFACTURING PROCESS ..................................2
1.3: THE PRODUCTS OF THE CHEMICAL INDUSTRY ...............................3
1.4: RAW MATERIALS ....................................................................4
1.5: SKILL AND SERVICE REQUIREMENTS FOR CHEMICAL MANUFACTURING.........8
1.6: CHEMICAL ENGINEERING PROJECTS ...........................................11
    1.6.1: New Chemical Engineering Projects ...........................11
    1.6.2: Existing Chemical Engineering Projects.......................12
    1.6.3: Project Components...............................................12
1.7: INFORMATION USUALLY REQUIRED FOR PROCESS ANALYSIS AND DESIGN ...........13
1.8: IMPORTANT CHEMICAL AND PHYSICAL DATA FOR PROCESS ANALYSIS ...............14
AND DESIGN ..............................................................................14
REFERENCES FOR CHAPTER ONE .......................................................15

CHAPTER TWO: PRE-INVESTMENT AND ECONOMIC EVALUATION OF
CHEMICAL ENGINEERING PROJECTS ..........................................17

2.1: THE INVESTMENT DECISION PROBLEM .........................................17
2.2: FACTORS WHICH AFFECT THE PROFITABILTY OF CHEMICAL ENGINEERING ............17
PROJECTS .................................................................................17
    2.2.1:  Capital Investment..............................................19
    2.2.2:  Manufacturing Expenses.......................................20
    2.2.3:  General and Administrative Expenses, $C_G$ ................21
2.3: FORECASTING FACTORS WHICH AFFECT THE PROFITABILTY OF .....................22
CHEMICAL ENGINEERING PROJECTS..................................................22
    2.3.1: Locating the Sources of Information...........................22
    2.3.2: Selecting the Cost Forecasting Method.......................24
2.4: ESTIMATING THE CAPITAL COSTS OF CHEMICAL PLANT EQUIPMENT ...............25
    2.4.1: Types of Capital Cost Estimates................................25
    2.4.2: Order of Magnitude Estimates ................................26
    2.4.3: Study Estimates .................................................35
    2.5.1: The Rule of Thumb or Ratio Method for Estimating Direct .........44
    Manufacturing Costs..................................................44
    2.5.2: The General Equation Method for Estimating Direct .................45
    Manufacturing Costs (Woods, 1975)..................................45
    2.5.3:  The Factor Method for Estimating Direct Manufacturing ..........46
    Costs (Peters & Timmerhaus, 1985) ..................................46
    2.5.4:  The Method of Varying Production Rates for Estimating ..........47

iv

*Direct Manufacturing Costs (Woods, 1975)*..............................*47*
*2.5.5: The Anergy Method for Estimating Direct Manufacturing*..........*49*
*Costs (Gaensslen, 1979)*................................................*49*
*2.5.6: Other Methods of Estimating Direct Manufacturing Costs* ........*49*
2.6: ESTIMATING GENERAL EXPENSES.............................................58
2.7: ESTIMATING SALES INCOME OR REVENUE......................................58
2.8: ESTIMATING THE PROFITABILITY OF PROPOSED INVESTMENTS.....................59
    *2.8.1:  Break -Even Analysis* .........................................*60*
    *2.8.2: The Return on Investment (ROI) Method* .........................*62*
    *2.8.3: Pay-out Time Method* ...........................................*64*
    *2.8.4: The Present Value Method*.......................................*66*
    *2.8.5: The Discounted Cash Flow Rate of Return Method*.................*67*
    *2.8.6: Method of Capitalised Costs*...................................*77*
2.9: EFFECT OF INFLATION ....................................................79
    *2.9.1: Effect of Inflation on Net Present Value (NPV)* ................*80*
    *2.9.2:  Effect of Inflation on DCFRR* .................................*82*
    *2.9.3: Effect of Inflation on Payback Period* .........................*82*
2.10: PROJECT RISK ANALYSIS .................................................82
    *2.10.1: Sensitivity Analysis*.........................................*83*
    *2.10.2: The Hurdle Rate Analysis*.....................................*83*
    *2.10.3:  Pay-off Table Analysis* .....................................*83*
    *2.10.4: The Mini-Max Loss Analysis*...................................*83*
    *2.10.5: Utility Function Analysis* ...................................*84*
    *2.10.6: Probability Analysis*.........................................*84*
    *2.10.7: The Monte Carlo Simulation*...................................*84*
REFERENCES FOR CHAPTER TWO .................................................85

**CHAPTER THREE: PROCESS DEVELOPMENT AND DESIGN**............................**87**

3.1: PROCESS DESIGN ........................................................88
3.2: PROCESS DESIGN STEPS ..................................................88
    *3.2.1: Flowsheets*....................................................*89*
    *3.2.2: Information required in all Flowsheets* .......................*90*
    *3.2.3: Types of Flowsheets* ..........................................*90*
3.3: MATERIAL AND ENERGY BALANCES..........................................101
3.4: EQUIPMENT DESIGN .....................................................101
    *3.4.1: Types of Equipment Drawings* .................................*102*
3.5: EQUIPMENT SPECIFICATIONS .............................................103
REFERENCES FOR CHAPTER THREE .............................................105

**CHAPTER FOUR: THE ELEMENTARY MATHEMATICS AND THERMODYNAMICS OF MATERIAL AND ENERGY BALANCES**.......................................**107**

4.1: MASS FRACTION........................................................107
4.2: MOLE FRACTION .......................................................108

4.3: Mass and Mole Ratios ..................................................................109
4.4: Molarity and Molality ...............................................................110
4.5: Pure Component Volume ...........................................................110
4.6: The Law of Amagat or Leduc's Law .........................................110
4.7: Ideal Gas Volume .....................................................................112
4.8: Partial Pressure ........................................................................112
4.9: Dalton's Law of Partial Pressures..............................................113
4.10: Concentration in the Gas Phase .............................................115
4.11: Density ....................................................................................115
  4.11.1: Density of Liquid Mixtures................................................116
  4.11.2: Density of Pure Gases......................................................117
4.12: Specific Heat Capacity.............................................................118
4.13: Enthalpy ...................................................................................124
4.14: The Second Law of Thermodynamics in Process Design...................125
4.15: Thermodynamic Descriptions of Phase Equilibrium.........................126
  4.15.1: Ideal Solutions and Mixtures ...........................................129
  4.15.2: Non-Ideal Solutions.........................................................134
  4.15.3: Prediction of Heats of Transition of Pure Substances.............153
  4.15.4: Prediction of Heats of Transition of Mixtures of Substances....163
4.16: Thermodynamic Descriptions of Chemical Equilibrium.....................167
  4.16.1: IUPAC Convention for Chemical Reaction Studies.................167
  4.16.2: Thermal Changes associated with Chemical Reactions..........169
  4.16.3: The Equilibrium Constant .................................................176
  4.16.4: Calculations involving Heats of Reaction, Formation,.............181
  Combustion etc. ........................................................................181
4.17: Other Methods of Predicting Phase. and Chemical Equilibrium ........187
  4.17.1: Predicting Fluid Phase Equilibrium....................................187
  4.17.2: Predicting Chemical Equilibrium .......................................193
  4.17.3: Predicting Chemical Reaction Rates..................................193

CHAPTER FIVE: MATERIAL AND ENERGY BALANCES.............................197

5.1: Basic Concepts...........................................................................197
  5.1.1: Definition of System Boundary ...........................................197
  5.1.2: Defining the Basis of Calculations......................................197
  5.1.3: Reacting and Non Reacting Systems...................................198
  5.1.4: Performance and Design Parameters ..................................198
  5.1.5: Methods for Material and Energy Balances .........................198
  5.1.6: Inventory, Assembly and Separation Operations...................200
  5.1.7: Basic Definitions of Process Units and Operations................202
5.2: Material Balances Across Individual Process Units..........................206
  5.2.1: Estimating Process Paramaters .........................................215
  5.2.2: Material Balance by the Method of Constraints....................217
  5.2.3: The Split Fraction Method ................................................252

vi

5.3: Energy Balances across Process Units ............................................. 260
    5.3.1: The Thermal Energy Balance ................................................... 261
    5.3.2: Application of the Thermal Energy Balance to a Process Unit ... 262
    5.3.3: Thermal Energy Balance by the Method of Constraints ........... 262
    5.3.4: Thermal Energy Balance by the Method of Split Fraction ......... 270
    Coefficients ................................................................................... 270
  REFERENCES FOR CHAPTER FIVE ...................................................... 278

CHAPTER SIX: THE PROCESS DESIGN REPORT .......................................... **279**

  REFERENCES FOR CHAPTER SIX ........................................................ 281

APPENDIX I: DENSITIES OF VARIOUS MATERIALS ..................................... **283**

APPENDIX II: DENSITIES OF VARIOUS LIQUIDS ......................................... **296**

APPENDIX III: COMMON PROPERTIES OF WATER ...................................... **297**

APPENDIX IV: THERMAL PROPERTIES OF STEAM ....................................... **299**

APPENDIX V: COMMON PROPERTIES OF AIR ............................................ **300**

APPENDIX VI: THE SPECIFIC HEATS OF SOME COMMON LIQUIDS AND
FLUIDS .................................................................................................. **301**

APPENDIX VII: THE SPECIFIC HEATS OF SOME COMMON GASES ............. **304**

APPENDIX VIII: TYPICAL VALUES OF EQUILIBRIUM CONSTANTS AND HEATS
OF REACTION ....................................................................................... **306**

APPENDIX XI: ENTHALPIES AND EQUILIBRIUM CONSTANTS OF FORMATION
OF SOME ORGANIC AND INORGANIC COMPOUNDS. ................................ **309**

  INDEX ................................................................................................. **311**

# CHAPTER ONE:
# FROM IDEA TO PROCESS TO PRODUCT

## *1.0: From Idea to Product*

Most products in the market place arose from ideas that were pursued to fruition. Everybody is full of ideas throughout one's life but only those ideas which come with some conception of a process and a strong desire for their implementation will lead to product.

In well established organizations, new products often arise from more efficient utilization of existing processes, from customer experience with existing products or from new demand based on new or different economic or social conditions.

In either case, the provision of product to the market place or to consumers involves, usually, one or more of the following

- A production or manufacturing process
- Production, manufacturing and packaging of the product
- selling the product to customers
- expecting a profit from these activities
- expecting to pay tax, whether excise or company tax
- ensuring that other people or the environment are not put at risk

All businesses, from the one person enterprise to the big conglomerates, need, at various levels and capabilities,

- some form of advertising, customer education, research and development (new and existing products)
- direction and management to formulate business policy, deal with tax issues and labour relations and problems
- co-ordination of the activities of many and diverse peoples and organizations which are encountered in the business.

## *1.1: Companies*

Business organizations are usually known as companies. These are corporate citizens which, by law, have to be registered or incorporated either as private companies or public limited liability companies. It is necessary to have them registered or incorporated for the following reasons:-

1

- for the protection of the consumers of their products
- for the protection of the owners of the company
- for reliable and well organised revenue collection by government or other civil authority

The types of company which may be registered or incorporated in Nigeria are

- Public Limited Companies (PLC)
- Private Limited Companies (LTD)
- Unlimited Companies (UNLTD)
- Companies Limited by Guarantee (LBG)
- Incorporated Trustees
- Business Names

Although the full details on the above can be found from the Nigerian Corporate Affairs Commission, it will be helpful to point out the following.

An incorporated company must have a registered *Memorandum of Association* which sets out all the business it is allowed to do. It must, also, have a registered *Articles of Association* which spells out how it carries out its administrative and executive decisions.

A *private company* is one whose shareholders or owners do not exceed fifty in number and whose shares are not transferable in public trading.

A *public company* is one whose shares are quoted in the stock exchange and are transferable only through public trading and by means of the rules set out for doing so at the stock exchange.

A *limited liability company* is one whose liabilities are limited. This means that if the company becomes bankrupt, the creditors may recover their debts only from the assets of the company but not from those of the shareholders or owners of the company. Both private and public companies can be limited.

### 1.2: The Chemical Manufacturing Process

The chemical manufacturing process can be likened to a box with inputs (raw materials and energy) and outputs (useful products, waste material and waste energy). It involves a number of consecutive operations or

2

steps which can be shown in a flow diagram which shows material streams. If there is no change of inventory within the plant, the sum of all inputs, by the law of mass and energy conservation, must equal the sum of all outputs. Using material and energy balances, the quantities of fuel and raw materials and the size and capacity of process units to produce a given quantity of products, can be determined.

Chemical manufacturing process operations can be either *unit operations* or *unit processes*. In unit operations, only physical processes are involved such as fluid flow, heat transfer, filtration, distillation, grinding, drying, evaporation, humidification, crystallization, gas absorption, solvent extraction. Unit operations, also, involve the assembly, separation or transfer of molecules. In unit processes, chemical change, in addition to physical processes, is, also, involved. Examples are nitration, chlorination, hydrogenation, oxidation, sulphonation, electrolysis, combustion, and polymerization.

Physical and Chemical Principles behind Unit Operations and Unit Processes of Chemical Engineering

Shreve (1984) and Perry & Green (1984) have summarized the various physical and chemical operations and activities which come under the general classifications of unit operations and unit processes. According to these classifications, certain principles are of a purely physical nature so that processing operations based on them are referred to as unit operations Processing operations which are based on principles of a mostly chemical nature and which lead to chemical conversion processes are called unit processes.

All unit operations and processes have their origins in the four main subjects of chemical engineering, namely, momentum transfer, heat transfer, mass transfer and chemical reaction and thermodynamics. Tables 1.1, 1.2, 1.3 and 1.4 list these physical principles, the unit operations arising from them, the chemical conversion principles and the unit processes which, also, arise from them.

## 1.3: The Products of the Chemical Industry

The chemical industry consists of the traditional industry and what is referred to as the allied industry. The traditional chemical industry performs those operations for which the industry is well known such as petroleum refining, the processing of chemical intermediates, the

manufacture of plastics and fibres and the production of key large
volume products such as sulphuric acid, ammonia, etc. The allied
industries consist of those other industries such as metal, textile, paper,
etc., which service or are serviced by the traditional chemical industry.

Products of the chemical industry are very many and number into
thousands. In addition, one product may give rise to a chain of others
which, in turn, can have their own chains. Certain key chemicals of the
chemical industry can, however, be identified and these are listed in
Table 1.5. These products, and others in their chains, are used directly
or indirectly in the manufacture of other products such as medicines,
fertilizers, radioactive materials, pesticides and disinfectants, cosmetics
and toiletries, dyes, explosives, etc.

Demand for these products, from any particular producer, is affected by
prices and the existence of rival manufacturers. Only those
manufacturers who optimise their use of resources, for the same quality
and price of product, continue to remain competitive. In the industry,
this means striking the right balance between capital expenditure,
manufacturing costs and selling and distribution costs for optimum sales
revenue.

### 1.4: Raw Materials

Raw materials are those materials which are in their natural,
unprocessed or untreated state. It is not anything that is used to make a
product. With very few exceptions, industrial products or consumer
goods are not made directly from raw materials. These products are
made from intermediates which can be primary, secondary, tertiary, etc.
derivatives of raw materials. Thus, to take a very simple example, iron
ore is a raw material which can be processed to ingots (or billets), a
primary intermediate. The billet can be cast or rolled to any shape such
as a rod or flat bar which becomes a secondary intermediate. This rod or
bar can form a final product such as a window protector or can become
a tertiary intermediate, such as a bolt, in another product and so on.

An important point to be clear about is about what is generally regarded
in Africa, south of the Sahara, as local when discussing raw materials.
The popular media and governments, in these countries, because certain
raw materials such as crude oil, natural gas, tin ore, columbite, etc, are,
mostly, exported in order to earn foreign exchange, do not regard them
as local raw materials. Instead, things like cassava, palm oil etc are

4

accepted to be local raw materials. The fact is that any untreated or
unprocessed material found in its natural state in any country is a local
raw material in that country. Table 1.6 lists some of the better known
raw materials.

**Table 1.1: The Physical Principles applicable to Industrial Chemical
Engineering Processes (Perry & Green, 1984)**

| S/No | Physical Principle | Application |
|------|--------------------|-------------|
| 1 | Fluid and particle mechanics | Fluid and particle flows and measurement |
| 2 | Transport and storage of fluids | Pumping and storage of liquids and gases by pumps, compressors, blowers, valves, piping, storage and process vessels |
| 3 | Handling of bulk and packaged solids | Conveying, packaging, storage |
| 4 | Size reduction and enlargement | Crushing, grinding, agglomeration, granulating, compacting |
| 5 | Heat generation and transport | Fuels, furnaces, combustion, power generation and transmission |
| 6 | Heat transmission | Heat transmission by conduction, convection and radiation |
| 7 | Heat transfer equipment | Evaporators, heat exchangers, etc. |
| 8 | Psychrometry, evaporative cooling, air conditioning and refrigeration | Water cooling, humidification, refrigeration and air conditioning |

**Table 1.2: The Unit Operations based on the Physical Principles of
Industrial Chemical Engineering (Perry & Green, 1984)**

| S/No | Unit Operation | Applications |
|------|----------------|--------------|
| 1 | Distillation | Binary and multicomponent distillation, extractive, azeotropic and molecular distillation |
| 2 | Gas absorption | Acid manufactures, purification of gas |

5

| | | |
|---|---|---|
| | | streams |
| 3 | Liquid extraction | Extractions and purifications |
| 4 | Adsorption & ion exchange | Extractions in solid/gas and solid/liquid systems |
| 5 | Miscellaneous separation processes | Leaching, crystallisation, sublimation, gaseous diffusion, dialysis, electrodialysis |
| 6 | Liquid - gas processing | Gas - liquid contacting, phase dispersion, phase separation |
| 7 | Liquid - solids processing | Filters and centrifuges, thickeners and clarifiers, paste mixers and agitators, ion exchange |
| 8 | Gas - gas, liquid - liquid, and solid - solid processing | Sampling, screening, froth flotation, electrostatic separation |
| 9 | Gas - solids processing | Contacting equipment for heat and mass transfer |

**Table 1.3: Chemical Conversion Principles applicable to Industrial Chemical Engineering Processes (Shreve & Brink, 1977)**

| S/No | Chemical Principle | Application |
|---|---|---|
| 1 | Acylation | Conversion of organic acid to acid halide |
| 2 | Alcoholysis | Conversion of acid halide to an ester |
| 3 | Alkylation | Insertion of an alkyl group into a ring or chain compound |
| 4 | Amination | To get an amide from an acyl halide |
| 5 | Ammonolysis | |
| 6 | Aromatization | Cyclization |
| 7 | Carboxylation | |
| 8 | Condensation | |
| 9 | Diazotisation | Diazonium salt from an aromatic amine |
| 10 | Esterification | |
| 11 | Friedel - Crafts | To add an alkyl group itno a ring |
| 12 | Halogenation | |
| 13 | Hydrogenation | |
| 14 | Hydrogenation/ de-hydrogenation and hydrogenolysis | |

6

| 15 | Nitration |
| 16 | Sulphonation |

**Table 1.4: Unit Processes based on Chemical Conversion Principles applicable to Industrial Chemical Engineering Processes (Shreve & Brink, 1977)**

| S/No. | Unit Process | S/No. | Unit Process |
|---|---|---|---|
| 1 | Calcination | 10 | Isomerisation |
| 2 | Causticization | 11 | Neutralization |
| 3 | Combustion | 12 | Controlled oxidation |
| 4 | Dehydration | 13 | Polymerisation |
| 5 | Double decomposition | 14 | Pyrolysis or cracking |
| 6 | Electrolysis | 15 | Reduction |
| 7 | Fermentation | 16 | Silicate formation |
| 8 | Hydrolysis and hydration | 17 | Sulphonation |
| 9 | Ion excahnge | | |

**Table 1.5: Key Products of the Chemical Industry (Nnolim, 1990)**

| S/No | Group | Key Chemicals |
|---|---|---|
| 1 | Oil based petrochemicals | Ethylene, Propylene, Butadiene, Benzene, p-Xylene |
| 2 | Natural gas based petrochemicals | Ammonia, Formaldehyde, Methanol, Urea |
| 3 | Chlor-alkalis | Caustic Soda, Chlorine, Soda Ash |
| 4 | Acids | Nitric acid, Phosphoric acid, Sulphuric acid |
| 5 | Industrial gases | Carbon Dioxide, Hydrogen, Nitrogen, Oxygen |
| 6 | Mineral based inorganics | Lime, Phosphorus, Potash, Sulphur |
| 7 | Plastic monomers | Styrene, Toluene Di-Isocyanate (TDI), Vinyl Chloride |
| 8 | Plastic polymers | Epoxies, Phenolics, Polyesters |

| 9 | Fiber monomers | Dimethyl Phthalate (DMT), Phthalic Acid (PTA) |
| 10 | Fiber polymers | Acrylics, Nylon 66, Polyesters |
| 11 | Adhesives and coating Monomers | Phenol, Vinyl Acetate |
| 12 | Pigments | Carbon Black, Titanium Dioxide |
| 13 | Solvents | Acetone, Methanol, Methylene Chloride |

**Table 1.6: Some Primary Raw materials**

| Minerals | Inorganics | Petroleum/organic | Agro-based |
|---|---|---|---|
| Alumina/bauxite | Limestone | Crude oil | Sugar cane |
| Cobalt | Gypsum | Natural gas | Maize |
| Zinc | Rock salt | Oil shale | Cassava |
| Iron ore | Asbestos | Tar sands | Sorghum |
| copper | | coal | Palm fruits |

## 1.5:  Skill and Service Requirements for Chemical Manufacturing

The chemical manufacturing process involves a number of skills and people of which the following are required. Not all companies can afford to retain these as employees, however. What skills a company or manufacturing process employs depend, almost entirely, on the products it makes.

<u>Research</u>

These come in four kinds namely:

*Basic research* seeks out new scientific principles or to gain new understanding of a known principle. Such research, for example, may be aimed at determining the dissolution rate of solid $CaCO_3$ in dilute acid, the kinetics and mechanism of a complex reaction or the mode of action of a new class of catalysts.

*Scouting research* aims to establish the technical feasibility of an idea. This may be, for example, to establish the technical feasibility of a mass catering system for the typical secondary or tertiary institution.

*Engineering researchs* concentrates on unit processes, unit operations

8

or system components to gain understanding of, or obtain data on, the behaviour or operation of these processes, operations or components. *Applications research* seeks to find new applications for existing or new products, processes and systems.

## Development

These can be of two kinds:-

*Product development* which gives birth to the evolution of a product. *Process development* which seeks to determine an economical and practical method of manufacturing an identified product.

## Process Engineering

Process engineering applies the disciplines such as unit operations, fluid flow, thermodynamics, heat and mass transfer, kinetics, economics, etc., to make the basic decisions required in the design of a new process.

## Project Engineering

This co-ordinates technical expertise such as research expertise, process design concepts, metallurgical constraints, structural requirements, process control and instrumentation, etc., and managerial skills such as materials procurement, work scheduling, logistical provisions, progress evaluation, etc., to translate technology into efficiently operating plants.

## Plant Operation

This ensures that the production or manufacturing plant runs efficiently, reliably and safely. It works to avoid mechanical failure, operator error, insufficient maintenance labour or spare parts, and manages production in adverse weather, labour disputes, raw material shortages, lack of distribution carriers and many other common problems likely to be encountered in a typical plant.

## Plant Engineering

This has the responsibility of keeping the mechanical components of the plant operating at top efficiency. It sees to the maintenance of equipment and facilities, utility operations (steam and electricity generation), the distribution of electricity, steam, water and compressed

9

air, the maintenance of stores, etc.

## Technical Management

This involves managing all technical professionals and programmes. To be in the technical management team, one has to have demonstrated sound knowledge and experience in their technical area together with a flair for managing and motivating other people.

## Technical and Business Consulting

No matter how good or knowledgeable a compnmay is in its business, there is always an area of the business in which the company will need more specialized advice or expertise, usually, from consultants who offer technical expertise as independent businessmen. Consultants are, often, people who have acquired some proficiency and experience in one or more aspects of the skills mentioned above.

## Government and Government Officials

Theoretically, government and government officials can only be involved in business in an advisory or regulatory capacity but sometimes they get involved in research depending on what the current government policy with regard to a particular industry is.

## Sales and Marketing

These are key to the success or failure of any manufacturing business. They provide real and effective communication between the manufacturer and its market by:-

- determining demand, quality and type of products required in the marketplace
- ensuring the effective sale of such products and services
- identifying new products or services not yet in existence which may need to be produced.

## Customer Technical Service

This is a department of the company which assists customers in the proper use of the company's products through offering technical advice and, in some cases, new product development or research assistance.

Lawyers

Lawyers are needed to draw up or scrutinize contracts and agreements entered into by companies or manufacturing entities and to represent their interests in case of litigation. It is wise to chose lawyers who are versed in patent law and are able to write, prosecute or defend, in a court of law, patent claims or infringement. They should, also, be competent in licensing, in negotiating or writing licensing agreements or in giving legal advice in engineering or R & D contracts, etc..

Accountants

These are the people who analyse the financial health of the business and prepare all the financial reports required by law and by shareholders of the company.

Other Skills

Other skill required outside those listed above will depend on the particular product and the services required.

## 1.6: *Chemical Engineering Projects*

A key element of industrial chemical engineering, especially in developing countries, is the execution of chemical engineering projects. A project is, by definition, a plan or scheme with definite objectives to be achieved. These objectives may be defined in terms of materials, in terms of cost, in terms of time or in terms of all of these combined.

A chemical engineering project is, therefore, any such project which is associated with changing the properties of some material in bulk. Like in all other engineering and non-engineering projects, any project can belong to any one of the two major classes of projects, namely:- new projects or existing projects requiring modification.

## 1.6.1: *New Chemical Engineering Projects*

A new project is regarded as a *grassroots project* if everything or almost everything concerning the new chemical plant is to be built from scratch. Such things include, not only the main chemical plant itself but also, all types of services and facilities such as access roads, water and electricity supplies, staff housing, schools, hospitals, etc.

11

A new project can, also, be *turn - key* if all the client, or the party for whom the plant was built, has to do is to, literally, switch on the plant by simply turning the key. Every item of chemical plant necessary for normal operation is installed, tested and commissioned so that client input is to simply pay for the plant and go into operation.

Grassroots and turn-key plants are more popular in developing countries where the municipal services or the technological base is deficient, inadequate or non-existent. These plants are, usually, more expensive than new plants for which the client has its own expertise in certain areas and, merely, purchases specific expertise, equipment or technology from those selling them.

Every new chemical engineering project, usually, consists of one or more of the following

- new process or product development,
- construction,
- start up.

### 1.6.2: *Existing Chemical Engineering Projects*

Often, improvement in demand of a company's products, the need to break into new markets or to increase market share, lead to the desire to expand existing production facilities. The expansion then becomes a project by itself.

A company may find that its products are no longer competitive in price or performance because its production plants or methods are outdated and need retrofitting or modernisation. Such retrofitting or modernisation then becomes a project.

### 1.6.3: *Project Components*

Chemical engineering projects, like all other projects, consist of the following components

- Management component, generally, referred to as project management
- Engineering/technical/scientific component, referred to as project engineering
- Cost component, increasingly, being referred to as cost engineering to differentiate it from mainstream accountancy, economics, and corporate and public finance, where detailed knowledge of engineering systems

12

is not a critical requirement.

## 1.7: *Information Usually Required for Process Analysis and Design*

The types of information required are information on

1. Manufacturing processes
2. Equipment parameters
3. Materials of construction
4. Costs
5. Physical and chemical properties of process material

Sources of Information

1. From company files, if process is a repeat project or similar to new one (restricted information)
2. From literature (experimental values or prediction methods)
3. From laboratory or pilot plant experiments and research

a). Common Literature Sources of Information on Manufacturing Processes

Information from the literature about manufacturing processes are, usually, promotional as most of relevant critical information is proprietary to the company that developed it. The more reliable sources are

i) Encyclopaedia of Chemical Technology by Kirk & Othmer
ii) Encyclopaedia of Chemical Processes and Design by McKetta
iii) Chemical Process Industries by Shreve & Brink
iv) Hydrocarbon Processing (Journal)
v) Patents

b) Common Literature Sources of Information on Physical and Chemical Properties

These are more factually accurate as they are based on experimental measurements and authenticated data.

i) International Critical Tables
ii) Tables and Graphs in Handbooks and Textbooks

13

iii)     Thermophysical Properties of Matter Data Service (TPRC) – Plenum Press

iv)      Engineering Sciences Data Unit (ESDU)

v)       Journal of Chemical Engineering Data

vi)      Chemical Abstracts (American Chemical Society)

vii)     Engineering Index (Engineering Index Inc., N.Y.)

viii)    Computerized Data Banks such as Physical Property Data Services of the British Institution of Chemical Engineers

## 1.8: Important Chemical and Physical Data for Process Analysis and Design

These are:

a)     Material Properties

       e.g., Density, heat capacity, viscosity, vapour pressure, solubility,

b)     System Properties

       e.g., Pressure, Volume, Temperature

c)     Process Properties

       e.g., Enthalpies, Heats of Solution, Vaporization, Reaction, etc

d)     Economic and Profitability Properties

       Costs, etc.

## References For Chapter One

1   Aris, R (1977): *Academic Chemical Engineering in Perspective*;
    Ind. Eng. Chem. Fundam.; Vol. 16, No. 1, pp 1-5; Am. Chem.
    Soc., Wash. D. C., USA

2   Barret, J. W. (1972): *Parallel Lines Must Meet*; The Chemical
    Engineer, No. 262; pp 215-220; Inst. Chem. Engrs., Rugby, UK

3   Bell, D (1976): *Welcome to the Post-Industrial Society*;
    ChemTech; Vol. 6, No. 10; pp 608-610; Am. Chem. Soc., Wash.
    DC., USA

4   Chemical Engineering (1978): *Wide World of Careers in
    Chemical Engineering*; Vol. 85, No. 1; pp 58-85; McGraw-Hill
    Book. Co., N.Y., USA

5   Danckwerts, P. V. (1972): *Chemical Engineering Science*; The
    Chemical Engineer, No. 262; pp 222-225; Inst. Chem. Engrs.,
    Rugby UK

6   Hougen, O. A. (1977): *Seven Decades of Chemical Engineering*;
    Chem. Eng. Prog.; pp 89-104; AIChE., N.Y.,USA

7   The Institution of Chemical Engineers (1977): *Regulations for
    the Election or Transfer within the Institution of Chemical
    Engineers*; Inst. Chem. Engrs., Rugby, UK

8   King, C. J.; West, A. S.(1976): *The Expanding Domain of
    Chemical Engineering*; Chem. Eng. Prog.; Vol. 72, No.3; pp 34-
    37; AIChE., N.Y., USA

9   Marvin, P (1973): *Fundamentals of Effective R&D Management*;
    ACS Audio Course; Am. Chem. Soc., Wash. D.C., USA

10  National Research Council (1987): *Frontiers in Chemical
    Engineering: Research Needs and Opportunities*; Committee on
    Chemical Science and Technology, Commission on Physical
    Sciences, Mathematics and Resources; National Research
    Council, USA

11  Nnolim, B. N. (1990): *Key Chemicals for the Nigerian Economy
    and Industry*; Proceedings of the AGM, NSChE, Lagos, Nigeria

12  Perry, R and Green, D, Editors: *Chemical Engineer's Handbook*;
    6th Edition; McGraw - Hill Book. Co., N.Y., USA (1984)

13  Shreve, R. N. And Brink, J. A. Jnr.; *Chemical Process*

*Industries;* 4th. Edition; McGraw - Hill Book. Co., N.Y,, USA (1977)

# CHAPTER TWO:
# PRE-INVESTMENT AND ECONOMIC EVALUATION OF CHEMICAL ENGINEERING PROJECTS

A chemical engineering project that is, technically, feasible needs to be, financially, viable if it is to be a sound business proposition. Such projects, therefore, need to be evaluated financially to determine their financial viability.

## 2.1: The Investment Decision Problem

Financial investments, whether in chemical engineering or in any other project, are expected to yield some profit. Profit can be interpreted differently by different people depending on their organisational, personal or social goals, A government will, for example, have a different view of profit from a public limited company. Philosophically, however, profitability is a measure of efficiency. Too much or too little profit vis-a-vis the investment can have undesirable universal consequences. Thus, it is not just enough to be profitable.  It is important that profitability be achieved under specified and acceptable conditions.

Once an acceptable definition of profit has been made, the issues which need to be resolved before deciding to invest or not to invest in the project are

- Identifying factors which affect the profitability of the investment
- Forecasting each of these factors
- Analysing the interplay of these factors and how they affect the profitability of the proposed investment.

## 2.2:  Factors which affect the Profitabilty of Chemical Engineering Projects

Table 2.1 outlines, in summary, the major factors which affect the profitability of chemical engineering projects.

For purposes of economic analysis, it is usual to group these factors under three main categories as

- capital investment

- manufacturing expenses
- general and administrative expenses.

Each of these is explained in further detail below.

**Table 2.1: Factors which affect the Profitabilty of Chemical Engineering Projects**

| S/No | Factor | Brief Explanation |
|------|--------|-------------------|
| 1 | Installed Cost of Fixed Equipment | |
| 2 | Working Capital | In cash or invested in inventories, often, a function of volume of business |
| 3 | Construction Period | It is a delay period in terms of earning income from the investment. If an on-going business is acquired, the construction period is zero. |
| 4 | Initial Start-up Expense | |
| 5 | Sales Volume and Product Price Forecast | |
| 6 | Cost stream throughout Project life | These costs include manufacturing costs and other overheads |
| 7 | Economic life of project | Or most likely period of successful operation. For manufacturing plants, life is between 10 and 15 years. |
| 8 | Effective depreciation life of depreciable fixed investment | This may be different from the economic life of the project. It is, usually, what is allowed by the government. |
| 9 | Salvage value of fixed investment | This is, often, assumed to be zero or equal to the cost of disposal. |
| 10 | Depreciation method applied | This affects the value of the cash flow. Three common methods are used - straight line method, declining balance method and sum-of- the-years digit method |
| 11 | Minimum acceptable rate of return on investment | Depends on company policy vis-a-vis the market, risks etc. |
| 12 | Income tax rate | Depends on government policy. Tax exemption/relief status of the project should be known |
| 13 | Inflation | Can be general or differential. Knowledge of appropriate indices is important. |

18

| 14 | Risk elements of the project | may depend on a) novelty of project b) nature of the business c) reliability of the sales forecast |
| 15 | General business climate | |

## 2.2.1: Capital Investment

This is the total amount of money required to procure the necessary equipment and facilities to start and operate the plant. It is made up of two items, namely, fixed capital and working capital. Items which make up the fixed and working capital are listed in Table 2.2.

**Table 2.2: Items which make up Capital Investment Costs**

| Fixed Capital | Working Capital |
|---|---|
| Direct Costs | Cash at hand |
| Purchased equipment cost Cost of installation of purchased equipment Installed cost of instrumentation and controls Piping (installed) Eleetricals (installed) Land Building and services Site improvements Service facilities (installed) | |
| Indirect Costs | Accounts Receivable |
| Engineering and Supervision Construction expense Contractor's fee Contingency | |
| | Money tied up in inventories of a) Raw materials and supplies carried in stock b) Finished products in stock c) Semi-finished goods in process |
| | Accounts payable |
| | Taxes payable |

19

## 2.2.2: *Manufacturing Expenses*

These are expenses connected with the manufacture of the product when the chemical plant is in operation. They consist of the direct production costs, the fixed charges and plant overhead costs. The items which make up these manufacturing costs are listed in Table 2.3. Some of these costs are, directly, attributable to the manufacture of the product, some only, indirectly. Some of these costs vary linearly with production rate, some not linearly while some are independent of production rate. All these are summarised in Table 2.4.

If the manufacturing costs, attributable directly to the manufacture of the product but varying, linearly, with production rate are designated by $C_L$, while those, directly, attributable but not varying, linearly, with production rate, by $C_R$ and those, directly, attributable but independent of production rate are represented by $C_I$, while those not, directly, attributable and also independent of production rate, by $C_O$, then the total manufacturing $C_M$ is given by

$$C_M = C_L + C_R + C_I + C_O \qquad (2.1)$$

**Table 2.3: Items which make up Manufacturing Costs**

| Direct Production Cost | Fixed Charges | Plant overheads |
|---|---|---|
| Raw materials | Rent | Medical |
| Utilities | Insurance | Safety and Protection |
| Packing, containers, shipping | Property taxes | General plant overheads |
| Royalties | Depreciation | Payroll |
| Operating labour | | Restaurant |
| Operating supervision | | Recreation |
| Maintenance and repairs | | Salvage |
| Plant and operating supplies | | Control laboratories |
| Catalysts and solvents | | Plant superintendence |
| | | Storage facilities |

**Table 2.4: Details of Manufacturing Cost Components**

| $C_L$ | $C_R$ | $C_I$: | $C_O$ |
|---|---|---|---|
| Items, directly attributable to product manufacture and varying, linearly, with production rate | Items, directly, attributable toproduct manufacture but not varying linearly with production rate | Items, directly, attributable to product manufacture but independent of production rate | Items not, directly, attributable to product manufacture and, also, independent of production rate |
| Raw materials | Operating labour | Rent | Medical |
| Utilities | Operating supervision | Insurance | Safety and protection |
| Packing and containers | Maintenance | Property taxes | General plant |
| Shipping | Plant and operating supplies | Depreciation | Payroll |
| Royalties (if not on a lump sum basis) | Laboratory analysis | | Restaurant |
| | | | Recreation |
| | | | Salvage |
| | | | Control laboratories |
| | | | Plant superintendence |
| | | | Storage facilities |

## 2.2.3: General and Administrative Expenses, $C_G$

These are expenses, other than capital and manufacturing costs, which are necessary for the company's operations. They are not, easily, attributable to any given product. They consist of

- administrative expenses
- distribution and marketing expenses
- research and development expenses
- cost of finance
- gross earnings expenses

21

Table 2.5 lists the details of general and administrative expense.

**Table 2.5: Details of General and Administrative Expenses, $C_G$.**

| Administrative Expenses | Distribution & Marketing Expenses | Research & Development Expenses | Financing Expenses | Gross Earnings Expenses |
|---|---|---|---|---|
| Executive wages | Materials handling | Salaries & Wages | Cost of borrowed funds | Expenses arising from income tax laws on gross earnings |
| Clerical Wages | Sales offices | Research facilities | | |
| Office Supplies | Sales men & women | Consultant Fees | | |
| Engineering expenses | Technical sales service | | | |
| Legal Expenses | Advertising | | | |
| Office Up-keep | | | | |
| Buildings | | | | |
| General communication | | | | |

## 2.3:  Forecasting Factors which affect the Profitabilty of Chemical Engineering Projects

The two most important steps in making forecasts is

- to locate the sources of technical, business and required information and
- to decide on the forecasting method to be used to give realistic forecast.

### 2.3.1: Locating the Sources of Information

A surprisingly large amount of information is available in published form and in all kinds of places. Some of the information may be out of date, some incomplete or too general. Yet they provide much valuable guide, especially, for preliminary cost estimates. Some of the more common sources of information are highlighted below.

i.    Price/Cost Indices

These are indices that record changes in composite material and labour costs, consumer and wholesale prices and, also, adequately, document past trends (Mascio, 1979). They answer general questions on trends of costs of raw materials, equipment or the particular product under consideration.

ii.    Institutional Publications

Examples of these are:- The United Nations Year Book; UNIDO publications on specific products; Federal Office of Statistics publications; Annual reports of the Central Bank of Nigeria, etc.

iii.    Buyer's Guides and Trade Directories

These answer questions on who makes or sells what, plus other relevant information. Examples of such guides are the Chemical Week Buyer's Guide, Mineral Yearbook, Metal Statistics. Modem Plastics Encyclopaedia, The Nigerian Trade Journal, The Thomas Register, Kemp's Directories, M.A.N. Directories., etc.

iv.    Abstracts and Indexes

These give reviews and analyses and, sometimes, detailed information about some industry or technology. Typical examples are the Chemical Abstracts, Applied Science and Technology Index, Engineering Index, Chemical Markets Abstracts, Business Periodicals Index, etc.

v.    Periodicals, especially. Trade Journals

These give special review and forecast editions. Typical examples are:- Chemical Week, Chemical and Engineering News, Adhesive Age, Modern Plastics. Hydrocarbon Processing, etc.

vi.    Publications of Trade and Professional Associations

These are professional, or trade position, papers and are sources of specialised information. Typical examples are the publications of the Sulphur Institute, the Lime Institute, the Manufacturers' Association of Nigeria, etc.

vii.      Handbooks. Reference Works, etc

These are, extremely, useful sources of technical and code of practice type of information. Examples are the Chemical Engineer's Handbook, Civil Engineer's Handbook, A.S.M.E. Pressure Vessel Code, etc.

## 2.3.2: Selecting the Cost Forecasting Method

Cost forecasting is, simply, the estimation of the financial investment that may be required for capital, manufacturing and general/administrative expenses for a plant or process. The method used will depend on stage at which the process is.

For example, where only the market price or desired annual production rate of a product is known, the methods used to make forecasts, of capital, manufacturing and other costs, are likely to be based on previous, general experience in the industry, with the same or similar products. Such methods are known as ratio or correlation methods since current costs are estimated in comparison to previous ones, on the basis that these two costs are related by some ratio or correlation.

Where a flow sheet has been developed, with material and energy balances done, the method used will be more detailed in order to account for the characteristics of the process units and conditions, which have been identified. In this case, factor methods are preferred. In this method, it is assumed that all costs are related to the fixed capital cost by a multiplying factor.

In the situation where each process unit is designed in detail, the costing effort and its accuracy will be high. All costs are obtained directly since the composition of any cost item is, accurately, known. Because detailed design of equipment is outside the scope of this book, this method is not included in further detailed discussion.

The methods for estimating depreciation, taxes, insurance, interest and investment costs are, fairly, standardised and do not depend on the stage of the project. Their accuracy depends, however, on the amount of information available on the project since more accurate estimates of

costs and revenue are associated with more accurate analyses which are
the results of advanced project analyses.

## 2.4: Estimating the Capital Costs of Chemical Plant Equipment

Estimating the cost of chemical plant equipment is, essentially, a job for
the professional cost engineers. Large companies, usually, have
departments which handle such costing. In smaller companies, however,
all costing may be handled by one or few individuals. In either case, the
practising chemical engineer needs to be aware of, and to understand,
the methods for predicting the costs of chemical plant equipment.

### 2.4.1: Types of Capital Cost Estimates

The accuracy of a cost estimate will vary depending on the reliability of
the cost data on which the estimate was based. Acceptable variability
for capital cost estimates of chemical plants are ± 40 % for an order of
magnitude estimate and ± 3 % for a detailed estimate. The various types
of estimates, in common use, are listed in Table 2.6.

**Table 2.6: Types of Estimates of Capital Investment (Pikulik & Diaz, 1977)**

| Type of Estimate | Usual Basis | Probable Accuracy, % |
|---|---|---|
| Order of Magnitude (Ratio) Estimate | Previous similar information | ±40 |
| Study (Factored) Estimate | Knowledge of flowsheet and major equipment | ±25 |
| Preliminary (Initial, Budget, Scope) Estimate | Sufficient data for budget preparation | ±12 |
| Definitive (Project Control) Estimate | Detailed data but not complete drawings | ±6 |
| Detailed (Firm, Contractor's) Estimate | Complete drawings and specifications | ±3 |

We shall limit our discussions to the order of magnitude and study
estimates because these are the only estimates for which a general
treatment of process development and analysis, as presented in this and

25

similar books, are applicable. There is, also, little or no indigenous data so that all data references will be made to the USA and Western European experience. Since most third world countries source their equipment from these regions, equipment costs estimated, using their cost data, are valid and only need conversion to local currency, using the prevailing currency exchange rates.

### 2.4.2: Order of Magnitude Estimates

Order of magnitude estimates can be obtained, using one or more of the four methods available. These methods are

- ratio methods
- cost and price index methods
- exponential methods
- Lang factor methods

1.    Ratio Methods

The quickest and most general method for generating order of magnitude estimates of capital costs is based on Capital Ratios, if these are available for the plant or equipment, under consideration. The capital ratio is concerned with integrated systems, not specific items, of plant. A major advantage of the method is that it does not depend on having a flow diagram of the process.

The capital ratio (CR) of a grass-roots plant is defined as the ratio of fixed capital investment, FC, to the annual sales revenue, AS (Lyn & Howland, 1960). That is

$$CR = \frac{FC}{AS}, \frac{Naira}{Naira\,per\,year} \qquad (2.2)$$

Capital ratios for a number of process industries are listed in Table 2.7.

Another rapid method, similar in effect to the capital ratio method, is the investment/capacity ratio method. It determines the ratio of capital investment per unit of annual production capacity for several process plants (Haselbarth, 1967). These are listed in Table 2.8. Grass-roots plants would cost 30 - 40% more while enlargements of existing plants would cost 20-30% less than the values given in the Tables.

Both the capital ratio and the investment/production capacity ratio

26

methods are, relatively, inaccurate and are useful, only, for preliminary screening of processes.

**Table 2.7: Capital Ratios for Selected Process Industries (Perry & Green, 1984)**

| Industry | Capital Ratio (1958) | Industry | Capital Ratio (1958) |
|---|---|---|---|
| General Chemicals | 2.02 | Pharmaceuticals | 0.92 |
| Carbon Black | 3.98 | Pigments, Paints and Inks | 1.04 |
| Glass | 1.46 | Pulp and Paper | 2.01 |
| Processed Foodstuffs | 0.66 | Resins and Plastics | 1.90 |
| Heavy Inorganics | 2.24 | Rubber | 1.04 |
| Petroleum | 3.08 | Soaps/Detergents | 0.69 |

**Table 2.8: Investment/Production Capacity Ratios for Process Plant (Perry & Green, 1984)**

| Product | Process Route | Capacity Range, (in 100 metric tons per annum) | Capital Cost per annual metric ton, S/tonne |
|---|---|---|---|
| Acetylene | Natural gas | 2- 150 | 1560 |
| Ammonia (anhydrous, liquified) | Steam reforming of natural gas. Partial oxidation of natural gas. Fuel oil. | 30 - 330 | 200 |
| | | 30 - 330 | 247 |
| | Coal. | 30 - 330 | 313 |
| | | 30 - 330 | 420 |
| Soap | | 1.5-7 | 289 |
| Sodium Hydroxide | Electrolysis of brine | 3-330 | 1110 |
| Sugar Cane | 120 days per year | 6-30 | 1070 |

*All costs are in North American dollars with M&S = 1000. It is assumed that there are 330 operating days per year.*

Another ratio method is the universal factor method (Woods, 1975). In this method, the total fixed capital cost, of either a grass-roots or a battery limits plant, is estimated from the present selling price of the product and the envisaged annual production capacity of the plant. The equation is

$$FC = \frac{S \times Q}{W} \tag{2.3}$$

where FC is the fixed capital cost, S the up to date selling price of the product, Q, the annual production rate of product, in consistent units, and W the universal factor or turn-over ratio. W has a value of 1 for most plants, 1.4 for large volume, organic, intermediates for which raw materials or labour costs dominate and falls within the range 0.2 to 8 in all cases.

The value of Q used has to be within the range for which the values of W were determined. Correlations exist for determining the selling price, S, but it is advisable to use only current market values. It is to be noted that, W, the turnover ratio is the reciprocal of the capital ratio. The annual sales, AS, is the product of the selling price, S, and the annual production rate, Q. Hence the capital ratio and the universal factor methods are, essentially, the same. Whichever one is used depends on the availability of data for its use.

**Illustrative Example 2.1**

It is desired to estimate the fixed capital cost of a vegetable oil processing plant using the Capital Ratio and Universal Factor methods. The plant is expected to produce 500,000 liters per year of refined oil. The average market selling price of the oil is N10.50 per liter.

**Answer**

From Perry & Green (1984), the capital ratio for foodstuffs processing plant is 0.66 years.

Annual sales,    $AS = 500{,}000 \times N10.50 = N5{,}250{,}000$
From equation (2.2),
$\qquad FC = CR \times AS = 0.66 \times N5{,}250{,}000 = N3{,}465{,}000$   *Ans.*
Using the Universal Factor method, equation (2.3),
$$FC = \frac{S \times Q}{W} = \frac{N10.50 \times 500{,}000}{1.515} = N3.465{,}347 \quad Ans$$

since   $S = N10.50$ per liter
$\qquad Q = 500{,}000$ liters per year

and $\quad W = \dfrac{1}{0.66} = 1.515 \; per \; year$

## 2. Use of Cost Indices

In estimating total installed costs of projects, whose completion and start-up dates may be a few years away, cost indices, when, properly, used, can provide a sound method for assessing the most likely cost variations over the life span of the project. Mascio (1979) produced, by regression analysis, long term and short term, trend equations for the more common indices applicable to the chemical industry in the U.S.A. Equations for other countries can be, similarly, derived if sufficient data are available.

For the U.S.A., long term trends were described over a seven year period by a linear equation of the form
$$P = \alpha + \beta t \qquad (2.4)$$
while short term trends, over a period of three years, were described by an exponential equation
$$P = \theta e^{\gamma t} \qquad (2.5)$$
P is the percent increase in the price index in year, $t$, based on the starting year, while $\alpha$, $\beta$, $\theta$, and $\gamma$ are constants. The labour cost indices were found to retain their linear form in both the long and short term. Table 2.9 lists some of the more used indices in the U.S.A.

**Table 2.9: Some of the more Common Indices in use in the U.S.A. (Mascio, 1979)**

| Composite Cost Indices | Material Cost Indices | Labour Cost & Price Indices |
|---|---|---|
| Engineering News Record Building Cost | Marshall & Swift Equipment Cost | Chemical Engineering Construction Labour |
| Dept. Of Commerce Construction Cost | Bureau of Labour Statistics, General Purpose Machinery & Equipment | Engineering News Record. Skilled Labour |
| Nelson Refinery Cost | Chemical Engineering Equipment & Machinery Support | Nelson Labour |
| Chemical Engineering Plant Cost | Nelson Materials | Engineering News Record. Common Labour |

29

| Bureau of Labour Statistics, Metal & Metal Products | Consumer Price Index | |
|---|---|---|
| | | Producer Price Index |

**Illustrative Example 2.2**

A piece of chemical engineering equipment cost $15,000 U.S. dollars in January, 1971. Determine the cost using both the long and short term Chemical Engineering Equipment cost index.

**Answer**

From the long term equation, for t = 7 years
$$P = \alpha + \beta t = -13.61 + 13.82t = -13.61 + 13.82 \times 7 = 83.13\%$$
Hence since
$Amount = principal + interest\ in\ period$
$$= principal\ (1 + interest\ rate) \qquad (2.6)$$
cost in January, 1978 = $15,000 (1 + 83.13/100) = $27.469.50 Ans

From the short term equation, for t = 3 years
$$P = \theta e^{\gamma t} = 49.32 e^{0.170 x 3} = 82.13\%$$
Cost in January, 1974 = $15,000 (1+82.13/100) = S27.319.5 Ans.

<u>Adjusting Indices to a Common Base</u>

Often, it is necessary to bring different indices to a common base year. With any year, the formula for conversion to a common base year is given by Mascio (1979) as

$$P_{over\ base\ year} = 100\left(\frac{published\ value}{conversion\ factor} - 1\right) \qquad (2.7)$$

where the published value is the present value of the index and the conversion factor is its value in the base year. Table 2.10 lists some of the conversion factors for the base year 1971.

3.   <u>Exponential Methods</u>

Another rapid method of estimating capital costs is the Capacity Ratio Exponent method, based on existing cost data. If the cost of a piece of equipment or plant, of size or capacity, $Q_1$, is $C_1$, then the cost of a similar piece of equipment or plant of size, $Q_2$, is given bv

$$C_2 = C_1 \left( \frac{Q_2}{Q_1} \right)^n \qquad (2.8)$$

$n$ is an exponent which depends on the type of equipment or plant. Although $n$ is, often, assumed to be 0.6 (the six tenth's rule), values of $n$, in practice, vary with type of equipment or plant.

**Table 2.10: Conversion Factors for Cost Indices (Mascio, 1979)**

| Composite Cost Indices | Conversion Factor | Actual Base Year (Index = 100) |
|---|---|---|
| Engineering News Record (ENR) Building | 875 | 1913 |
| Dept of Commerce Construction | 125 | 1967 |
| Nelson Refinery | 387.2 | 1946 |
| Chemical Engineering Plant | 128.2 | 1957-59 |
| **Material Cost Indices** | | |
| Marshall & Swift (M&S) Equipment | 315.2 | 1926 |
| Bureau of Labour Statistics (BLS) Equipment | 117.0 | 1967 |
| Chemical Engineering (CE) Equipment | 125.7 | 1957-59 |
| Nelson Materials | 257.9 | 1946 |
| BLS Metal | 116.5 | 1946 |
| **Labour Cost Indices** | | |
| CE Coastrnction Labour | 142.5 | 1957-59 |
| ENR Skilled Labour | 1,411 | 1913 |
| Nelson Labour | 473.4 | 1946 |
| ENR Common I-abour | 2,963 | 1913 |
| Producer Price Index | 118.8 | 1967 |
| Consumer Price Index | 119.2 | 1967 |

Tables 2.11 and 2.12 list various values of $n$, $Q_1$ and $C_1$ for selected plant and individual equipment, respectively. For process plants, capacity is expressed as metric tons per year while, for individual equipment, it is expressed in terms of a parameter related to its capacity such as surface area for heat exchangers, horse power for pumps, etc.

31

Also, for individual equipment, the cost may be F.O.B. (free on board), C.I.F. (Cost, Insurance, Freight) or INST (Installed) so that it is necessary to be sure which one is being used.

It is, often, advisable to update the cost, estimated from equation (2.8) using the appropriate equipment cost index. If $I_2$ is the index in the year for which an up-dated $C_2$ is sought and $I_1$ is the index in the year for which $C_1$, is obtained, then the updated cost is given by

$$Updated\,C_2 = C_2 \; from \; equation(2.8).x.\left(\frac{I_2}{I_1}\right) \qquad (2.9)$$

**Table 2.11: Typical Exponents for Estimating Capital Costs of Process Plants (Perry & Green, 1984)**

| Product | Process Route | Capacity, 1000 metric tons/yr | Approx. Cost, $ million | Capacity Range, 1000 tonnes | Fxponent, n |
|---------|---------------|-------------------------------|-------------------------|-----------------------------|-------------|
| Acetic Acid | Methanol | 10 | 77 | 2-50 | 0.45 |
| Acetylene | Natural Gas | 15 | 23.4 | 2-150 | 0.70 |
| Ammonia | Steam reforming of natural gas | 150 | 30 | 30 - 330 | 0.70 |
| | Partial oxidation of natural gas | 150 | 37 | 30 - 330 | 0.70 |
| | Fuel oil | 150 | 47 | 30 - 330 | 0.77 |
| | Coal | 150 | 63 | 30 - 330 | 0.77 |
| Soap | | 3 | 0.87 | 1.5-7.0 | 0.23 |
| Sodium Hydroxide | Electrolysis of brine | 30 | 33.4 | 3-300 | 0.38 |
| Sugar | 120 days per year | 10 | 10.7 | **6-30** | 0.41 |

## Illustrative Example 2.3

A centrifugal blower, operating at 27.6 $kN/m^2$ pressure, is to be purchased to deliver 40 standard $m^3/s$. Estimate the delivered cost of this blower. Note: the drive motor is not included.

## Answer

We shall use the exponent method, $C_2 = C_1 \left( V_2 / V_1 \right)^n$ where $C_1$, $V_1$ are the cost and volume discharge of a previous blower and $C_2$, $V_2$ are the cost and volume discharge of the current blower. From Table 2.12, $V_1 = 4.72$ standard $m^3/s$. $C_1 = \$67,000$. Since $V_2 = 40$ standard $m^3/s$

$$C_2 = \$67,000 \left( \frac{40}{4.72} \right)^{0.6} = \$241,515.83 \quad Ans$$

### Table 2.12:  Typical Exponents for Estimating Capital Costs of Individual Equipment (Perry & Green. 1984)

| Equipment | Size | Unit of Capacity | Approx. Cost, $1000 | Size Range | Exponent, $n$ |
|---|---|---|---|---|---|
| Agitator, turbine, top entry, open (FOB; | 10 (7.5) | Hp (kW) | 7.0 . | 2-30 (1.5-22.4) | 0.45 |
| Blower (28 $kN/m^2$). (centrifugal), (delivered) excluding motor | 10 (4.72) | $10^3$ std cuft/min ($m^3/s$) | 67 | 0.24-71 | 0.60 |
| Centrifugal     Pump (FOB),     excluding motor | 10(7.5) | Hp (kW) | 1.6 | 0.5 – 40 (0.37-30) | 0.30 |
| Conveyor     Screw (delivered),  excluding motor | 70 (540) | ft x m (m x mm) | 10 | 50-100 (390 - 780) | 0.46 |
| Dryer, Vacuum, Shelf (FOB), excluding trays, vacuum equipment | 9.3 | $m^3$ | 17 | 1.4-93 | 0.56 |
| Dryer, Drum (FOB), excluding motor | 9.3 | $m^2$ | 73 | 0.9 - 37 | 0.52 |

Four sources of error are associated with this method (Perry & Green, 1984), namely:-

- Using only one variable to correlate the cost of an equipment is an over-simplification
- So, also, is the representation of data by a single exponent

33

- Effects of technological improvements in equipment costing, $C_2$, of capacity, $Q_2$, since the time equipment of capacity, $Q_1$, was costed at $C_1$, are often ignored
- Errors, introduced by special circumstances, reduce the similarity between equipment with capacities $Q_1$ and $Q_2$

## 4. Use of Lang Factors

In this method, the fixed capital cost, FC, for either a grass-roots or battery limits plant, as well as the total capital cost for major additions to existing plant, can be estimated by multiplying the base cost by a factor, known as the Lang factor, $f_L$. The base cost may be on the basis of FOB, delivered or installed cost, for major equipment. That is

$$FC = f_L . EC \qquad (2.10)$$

where EC = base equipment cost.

Several authors {Wood (1975), Peters & Timmerhaus (1985), Coulson, Richardson & Sinnott (1984)} give values of $f_L$ for, predominantly, solids, fluids and mixed solids - fluids processing plants. Typical values are listed in Table 2.13.

Care must be taken, in using these factors, to determine whether the values given are for grassroots or battery limits plant or for major addition to existing plant. It should, also, be determined whether the equipment cost is FOB, delivered or installed and whether these costs include or exclude the cost of land, site preparations, plant auxiliaries and contractor's fees.

**Table 2.13: Lang Factors for Estimating Capital Costs**

| Type of Plant | Fixed Capital Cost<br>Woods (1975)[3] | Fixed Capital Cost<br>Peters & Timmerhaus (1985)[4] | Fixed Capital Cost[1]<br>Coulson & Richardson (1984)[2] | Fixed Capital Cost[5]<br>Woods (1975)[6] |
|---|---|---|---|---|
| | 3.8 | 3.9 | 3.1 | 2.2 |
| | 4.1 | 4.1 | 3.6 | 2.5 |
| | 4.8 | 4.8 | 4.7 | 3.3 |

Notes

1    Based on delivered equipment cost

| 2 | The original factors quoted by Lang |
| 3 | Based on extensive study of battery limits plants. For grass - roots plants, $f_L = 4.3$ for mixed solids-fluids processing plant. Does not include the cost of land. |
| 4 | Includes cost of land and contractor's fees |
| 5 | Based on installed equipment cost |
| 6 | Battery limits plant |

### *2.4.3: Study Estimates*

Study estimates are improvements on the order of magnitude estimates and represent attempts to estimate, more accurately, the cost of chemical plant. They are, particularly, useful when there are no detailed information on equipment designs for the process. The methods used are, often, refinements of factor methods or cost correlations. Some of these are described below.

1.      Method of Percent of Delivered Cost

More accurate estimates of fixed capital cost can be obtained if the various component costs of fixed capital cost are expressed as a percentage of the delivered equipment cost. The fixed capital cost FC is, then, given by

$$FC = DEC . \sum_i f_i + \left( DEC . \sum_i f_i \right) . \sum_j f_j \qquad (2.11)$$

where DEC = delivered equipment cost,
$f_i$ = multiplying factors for direct cost items such as electricals, installation, etc,
$f_j$ = multiplying factors for indirect cost items such as design, contractor's fees. etc.

Typical values of $f_i$ and $f_j$ are listed in Tables 2.14 and 2.15 while Peters & Timmerhaus (1985) and Woods (1975) give more extensive tables of values.

Thus, direct costs, DC is given by
$$DC = DEC . (1 + f_1 + f_2 + f_3 + ... f_9) \qquad (2.12)$$
and fixed capital cost. FC, becomes
$$FC = DC . (1 + f_{10} + f_{11} + f_{12}) \qquad (2.13)$$
Thus the total fixed capital investment (TFC) is related to the total direct costs (TDC) and the total indirect costs (TIDC) as

$$TFC = TDC + TIDC \qquad (2.14)$$

The difference between Tables 2.14 and 2.15 lies in the fact that in Table 2.14, the total fixed cost is known and the direct and indirect costs may be estimated using the appropriate factors while in Table 2.15, estimation of the indirect cost depends on a knowledge of the direct cost and both depend on a knowledge of the delivered equipment cost.

**Table 2.14: Typical Percentages of Delivered Equipment Costs for Multi-purpose Plants (Peters & Timmerhaus, 1985)**

| Direct Costs | Percent, Range, % | Indirect Costs | Percent, Range, % |
|---|---|---|---|
| Purchased Equipment | 15-40 | Engineering & Supervision | 4-21 |
| Installation of Purchased Equipment | 6-14 | Construction Expense | 4-16 |
| Instrumentation & Controls (Installed) | 2-8 | Contractor's Fee | 2-6 |
| Piping (Installed) | 3-20 | Contingency | 5-15 |
| Electrical (Installed) | 2-10 | | |
| Buildings (including services) | 3-18 | | |
| Site Improvements Service Facilities (Installed) | 2-5 8-20 | | |
| Land | 1-2 | | |

## 2.    Improved Lang Factors

The accuracy of the Lang factor method can be improved by decomposing it into several other factors for greater accuracy of estimates. Peters & Timmerhaus (1985) report the following factors:- 4.0 for fractionating columns, pressure vessels, pumps and instruments, 3.5 for heat exchangers, 2.5 for compressors and 2.0 for fired heaters.

In another approach, the Lang factors are decomposed into three other factors, namely:

$f_F$ = factor for cost of field labour
$f_P$ = factor for cost of piping materials
$f_M$ = factor for cost of miscellaneous materials such as instruments, insulation, foundations, buildings, etc.

Then, for a battery limits plant (Woods, 1975)

$$FC = PEC.(1 + f_F + f_P + f_M) + IEC + CRA \quad (2.15)$$

for a fluids processing plant, and

$$FC = PEC.(1 + f_M) + [PEC.(f_F + f_P)]_{fluid}$$
$$+ [PEC.(0.65 f_F)]_{solids} + IEC + CRA \quad (2.16)$$

for a fluids-solids processing plant.

PEC is the purchased equipment cost on FOB basis, IEC is the installed equipment cost and CRA is the incremental cost of corrosion resistant alloy materials.

**Table 2.15:  Typical Percentages of Fixed Capital Investment for Multi-purpose Plants (Coulson, Richardson & Sinnott, 1983)**

| Item | Fluids Processing | Solids-Fluids Processing | Solids Processing |
|---|---|---|---|
| **1. Direct Costs** | | | |
| Major equipment. | 1 | 1 | 1 |
| Installation, $f_1$ | 0.40 | 0.45 | 0.50 |
| Piping, $f_2$ | 0.70 | 0.45 | 0.20 |
| Instrumentation, $f_3$ | 0.20 | 0.15 | 0.10 |
| Electrical, $f_4$ | 0.10 | 0.10 | 0.10 |
| Process buildings, $f_5$ | 0.15 | 0.10 | 0.05 |
| Utilities, $f_6$ | 0.50 | 0.45 | 0.25 |
| Storage , $f_7$ | 0.15 | 0.20 | 0.25 |
| Site development, $f_8$ | 0.05 | 0.05 | 0.05 |
| Ancillary buildings $f_9$, | 0.15 | 0.20 | 0.30 |
| **Sub-total** | **3.40** | **3.15** | **2.80** |
| **2. Indirect Costs** | | | |
| Design & Engineering, $f_{10}$ | 0.30 | 0.25 | 0.20 |
| Contractor's fee, $f_{11}$ | 0.05 | 0.05 | 0.05 |
| Contingency, $f_{12}$ | 0.10 | 0.10 | 0.10 |
| **Sub-total** | **0.45** | **0.40** | **0.35** |

The material and labour factors are given by Hirsch and Glazier (1960) [in Peters and Timmerhaus, (1985)] as

$$\log_{10} f_F = 0.635 - 0.154.\log_{10}(0.001\,PEC) - 0.992\frac{TEC}{PEC} + 0.506\frac{f_V}{PEC} \quad (2.17)$$

$$\log_{10} f_P = 0.266 - 0.014.\log_{10}(0.001\,PEC) - 0.156\frac{TEC}{PEC} + 0.556\frac{TPC}{PEC} \quad (2.18)$$

$$\log_{10} f_M = 0.344 + 0.033.\log_{10}(0.001\, PEC) + 1.194\frac{TSC}{PEC} \qquad (2.19)$$

where

$TEC =$ total heat exchanger cost less incremental cost of alloys

$f_v =$ total cost of field fabricated vessels less incremental cost of alloys

$TPC =$ total pump plus driver cost less incremental cost of alloys

$TSC =$ total cost of tower shells less incremental cost of alloys.

## Illustrative Example 2.4

It is desired to estimate the fixed capital cost of a, predominantly, fluids processing plant. Study estimates indicate that, in this plant, heat exchangers, in carbon steel, would cost about N600,000; field fabricated vessels about N450,000; pumps, plus drivers, N550,000 and towers, N700,000. It is estimated that the incremental cost of necessary alloying of equipment will amount to N750,000. Determine the fixed capital cost of the plant.

## Answer

Using the improved Lang factors of 4.0 for fractionating columns, pressure vessels, pumps and instruments and 3.5 for heat exchangers, the fixed capital cost, FC, is, from equation (2.10)

$$FC = 4.(N700,000 + N550,000 + N450,000) + N750,000 + 3.5\, x\, N600,000$$
$$= N9,650,000 \quad Ans$$

Using the method of Hirsch & Glazier (1960), from equation (2.15)
$$FC = PEC.(1 + f_F + f_P + f_M) + IEC + CRA \qquad from\,(2.15)$$
The purchased equipment cost, PEC, is
$$PEC = N600,000 + N450,000 + N550,000 + N700,000 = 2,300,000 \quad (1)$$

$$\log_{10} f_F = 0.635 - 0.154.\log_{10}(0.001\, PEC) - 0.992\frac{TEC}{PEC} + 0.506\frac{f_V}{PEC}$$

$$= 0.635 - 0.154.\log_{10}(0.001\, x\, 2,300,000) - 0.992\frac{600,000}{2,300,000} + 0.506\frac{450,000}{2,300,000}$$

$$= 0.635 - 0.154 x 3.3617 - 0.992 x 0.2609 + 0.506 x 0.1957 = -0.043$$

That is
$$f_F = 0.9068 \qquad (2)$$

$$\log_{10} f_P = 0.266 - 0.014 . \log_{10}(0.001\, PEC) - 0.156 \frac{TEC}{PEC} + 0.556 \frac{TPC}{PEC}$$

$$= 0.266 - 0.014 . \log_{10}(0.001 \times 2,300,000) - 0.156 \frac{600,000}{2,300,000} + 0.556 \frac{550,000}{2,300,000}$$

$$= 0.266 - 0.014 \times 3.3617 - 0.156 \times 0.2609 + 0.556 \times 0.2391 = 0.3112$$

That is

$$f_P = 2.0473 \tag{3}$$

$$\log_{10} f_M = 0.344 + 0.033 . \log_{10}(0.001\, PEC) + 1.194 \frac{TSC}{PEC}$$

$$= 0.344 + 0.033 . \log_{10}(0.001 \times 2,300,000) + 1.194 . \frac{700,000}{2,300,000}$$

$$= 0.344 + 0.033 \times 3.3617 + 1.194 \times 0.3043 = 0.8183$$

That is

$$f_M = 6.5807 \tag{4}$$

But IEC = 0 and CRA = N750,000 hence

$$FC = PEC . (1 + f_F + f_P + f_M) + IEC + CRA \qquad from\ (2.15)$$
$$= N2,300,000 . (1 + 0.9068 + 2.0473 + 6.5807) + 0 + N750,000 .$$
$$= N24,980,040 \quad Ans$$

Note that this is more than twice the amount estimated using improved but lumped Lang factors.

## 3.  Improved Exponent Methods

An improvement on the power factor (six tenths factor) method, which involves both direct and indirect plant cost, is expressed as

$$FC = f(DC . R^n + IC) \tag{2.20}$$

where $f$ is the lumped cost index factor, relative to the original installation cost, DC is the direct cost of the originally installed equipment, IC is the indirect cost of the originally installed equipment, $n = 0.6$, the six tenths factor, and R is the ratio of process capacity of the new plant to the original installation.

The lumped cost index factor, $f$, is given by

$$f = f_E . f_L . \varepsilon_L \tag{2.21}$$

where $f_E$ is the current equipment cost index relative to the cost of purchased equipment, $f_L$ is the current labour cost index in the new location relative to purchased equipment labour cost and labour-employee hours for specific materials and $\varepsilon_L$ is the labour efficiency index in the new location relative to purchased equipment labour cost and labour-employee hours for specific materials,

## Illustrative Example 2.5

A process plant was erected in 1972 in the New England area of USA for $560,000. Jt is desired to estimate the fixed capital investment in a similar plant in 1976 of thrice the capacity and in the Pacific Coast of the same USA. The Marshall & Swift Installed Equipment cost index for the industry was 332 in 1972 and 479 in 1976.

The relative labour rates in New England and the Pacific Coast were 0.90 and 1.04, respectively, while the relative productivity factors were 0.85 in New England and 0.81 in the Pacific Coast.

### Answer

From the information given, with regard to equations (2.20) and (2.21), $R = 3$, $f_E = 479/332 = 1.44$, $f_L = 1.04/0.90 = 1.16$ and $\varepsilon_L = 0.81/0.85 = 0.95$. Since the fixed capital investment in 1972 was $560,000, the direct and indirect fixed costs can be estimated in 1972 using the method of percent of installed equipment costs as follows.

A table of direct and indirect costs is developed as shown below

| Direct Cost item | Assumed % of total cost | Cost, $ | Indirect Cost item | Assumed % of total cost | Cost, $ |
|---|---|---|---|---|---|
| Purchased equipment | 25 | 140,000 | Engineering & supervision | 5 | 28,000 |
| Purchased equipment installation | 10 | 56,000 | Construction expense | 10 | 56,000 |
| Instrumentation & controls (installed) | 5 | 28,000 | Contractors' fee | 3 | 16,800 |
| Piping (installed) | 10 | 56,000 | contingency | 5 | 28,000 |
| Electrical (installed) | 5 | 28,000 | | | |
| Building (including services) | 10 | 56,000 | | | |
| Site | 3 | 16,000 | | | |

| improvements | | | | | |
|---|---|---|---|---|---|
| Service facilities (installed) | 8 | 44,800 | | | |
| Land | 1 | 5,600 | | | |
| Total direct cost | 77 | 431,200 | Total indirect cost | 23 | 128,800 |

Total Fixed Costs = total direct costs + total indirect costs

$$= \$431{,}200 + \$128{,}800 = \$560{,}000. \qquad (1)$$

From equation (2.21)

$$f = f_E \cdot f_L \cdot \varepsilon_L = 1.44 \, x \, 1.16 \, x \, 0.95 = 1.59 \qquad (2)$$

Using equation (2.20)

$$FC = f\left(DC \cdot R^n + IC\right) = 1.59\left(\$431{,}200 \, x \, 3^{0.6} + \$128{,}800\right) = \$1{,}530{,}197.08 \; Ans.$$

## 4.    Anergy Method

An interesting method, based on anergy, which yields estimates of capital investment with +25 % accuracy, has been reported (Gaensslen, 1979). Anergy is that part of energy which can, no longer, in a given situation, be converted to useful work.

The basis of the method is that the investment cost of chemical plants (all using organic raw materials) was found, by regression analysis, to be related to the energy transformation in their processes and independent of the internal structure of the plant. Specifically, the process dependent investment cost, IC, was found to be, directly, proportional to the annual dissipated energy. That is

$$IC = k.DE \qquad (2.22)$$

where DE is the dissipated energy and k is a proportionality constant. If the investment cost and the dissipated energy are expressed per unit of production, then

$$IC = \frac{k \cdot LHV \cdot (1 - \alpha)}{\alpha} \qquad (2.23)$$

where LHV is the lower heating value of the product or products and $\alpha$ is the overall thermal efficiency of the process, based on enthalpy.

Equation (2.23) is valid for large capacity plants, with complex production processes, based on organic raw materials and which involve substantial energy transformations. In addition, the plant should have today's level of thermal optimisation, with the bulk of its dissipated energy leaving through equipment walls. The plant should, also, have an average mix of equipment, materials and catalysts.

Table 2.16 lists some $k$ values determined for chemical plants in Germany. The data required for the use of equation (2.23) are

- The lower heating value of all input materials, fuel gases, output products, output by-products and output fuel gases. These are, relatively, easy to obtain.
- Energy input from steam ($2.095 \times 10^6$ kJ/tonne) and electricity (3603 kJ/kWh). Energy output from excess steam is not considered.

**Table 2.16: Values of $k$ for Various Chemical Plants (Gaensslen, 1979)**

| Product | Raw Material | Specific Investment, DM/tonne per annum | Specific Dissipated Energy, $10^9$ gJ/tonne | $k$, DM/gJ per annum |
|---|---|---|---|---|
| Ethylene | $C_3/C_4$ | 554 | 22.2 | 24.95 |
| Hydrogen Cyanide | $CH_4/NH_3$ | 604 | 26.8 | 22.54 |
| Acetic Acid | $CO/CH_3OH$ | 241 | 10.0 | 24.10 |
| Acetylene | Naphtha | 2160 | 86.3 | 25.03 |
| Acetylene | $CH_4$ | 2160 | 103.0 | 20.98 |
| Syngas | Bituminous coal | $192/1000 \ m^3$ | 8.4 | 22.86 |

**Illustrative Example 2.6**

Determine the investment cost of a plant which uses 5 million metric tons of coal, 8000 kWh per annum of electrical energy and 2 million tons per annum of steam to produce 3 million metric tons of an organic product which has a LHV of 25,140 kJ/kg. The anergy constant is 22 DM per gJ per annum.

**Answer**

The LHV of coal is estimated to be 31,425 kJ/kg

Annual Input of Energy

Energy from coal = $5 \times 10^9$ kg $\times$ 31,425 $\times$ $10^3$ J/kg = 1.57125 $x$

42

$10^{17}$J/annum

Electrical energy = 3603 kJ/kWh x 8000 kWh/annum = 2.8824 x $10^{10}$ J/annum

Steam = 2 x $10^6$ tons x 2.095 x $10^9$J/tonne = 4.19 x $10^{15}$ J/annum

Total energy input per tonne of product

$$= \frac{1.57125x10^{17} + 2.8824x10^{10} + 4.19x10^{15}}{3x10^6}, \frac{J}{annum}. \frac{annum}{metric\ tons\ product}$$
$$= 53.77168 gJ/metric\ ton\ product \qquad (1)$$

Annual Output of Energy

Energy of product
$$= \frac{3x10^9 \ x \ 25140x10^3}{3x10^6}, \frac{kg\ product}{annum}. \frac{J}{kg\ product}. \frac{annum}{metric\ tons\ product}$$
$$= 25.140 gJ/metric\ ton\ product \qquad (2)$$

Annual Dissipated Energy, DE

DE = total energy input – total energy of product
$$= 53.77168 - 25.14 = 28.63168 gJ / metric\ tons\ product \qquad (3)$$

Investment Cost, IC

$$IC = k\ DE = 22\ x\ 28.63168x3x10^6 \frac{DM}{gJ/annum}. \frac{gJ}{tonne}. \frac{tonne}{annum}$$
$$= 1889.69x10^6\ DM \quad Ans.$$

5.     Refinement of the Anergy Method

Further refinements of the assumptions relating, specifically, to the internal efficiency and technical maturity of the plant lead to the equations

$$IC = LHV.a.\left(b + \frac{\alpha_P}{1-\alpha_P}\right).\left(\frac{1-\alpha_T.\alpha_P}{\alpha_T.\alpha_P}\right) \qquad (2.24)$$

where $a$ and $b$ are constants and $\alpha_T$ is the theoretical, $\alpha_P$ the practical, efficiency of the plant. In Germany, in 1977, $a$ and $b$ were found to be 9.3 DM/ gJ per annum and 0.15, respectively.

## 2.5: Estimating Manufacturing Costs

Manufacturing cost consists, essentially, of two major components. These are
- direct production cost
- indirect production cost, often, referred to as plant overheads

The use of the terms, direct and indirect, arises from the fact that those costs which are, clearly, attributable to the manufacturing of the product are the direct costs while those not, clearly, attributable are the indirect costs.

Direct production costs are of three types, namely

- those which vary, linearly, with production rate
- those which vary, but not linearly, with production rate and, in addition, are not zero at zero production rate
- those which do not vary with production rate because they are independent of production rate. They are, often, referred to as fixed costs or by other similar names.

Table 2.17 lists these components of manufacturing cost. Sections 2.5.1 to 2.5.6 describe the various methods in use for estimating the various kinds of manufacturing costs.

## 2.5.1: The Rule of Thumb or Ratio Method for Estimating Direct Manufacturing Costs

It has, somehow, become common wisdom that it is not worth making a product if its direct manufacturing cost per unit is greater than half its market selling price per unit. That is

$$Manufacturing\ cost\ per\ unit = 0.5\ x\ Selling\ price\ per\ unit \qquad (2.25)$$

Further experience in evaluating this rule of thumb suggests that 0.6 instead of 0.5 should be used in equation (2.25). This method, however, cannot be more than a rule of thumb, especially, in the manufacture of products such as organic chemicals where raw material cost, alone, can be up to 75% of the selling price of product.

44

## 2.5.2: The General Equation Method for Estimating Direct Manufacturing Costs (Woods, 1975)

This method assumes, first of all, that there are four elements of manufacturing cost, namely, raw material costs, cost of utilities, investment costs and labour costs. Secondly, it assumes that these costs are related in a linear manner. Thus, if R represents the raw material cost per unit of product, U the utility cost per unit of product, L the labour cost per unit of product and I the total fixed investment, the total manufacturing cost, MC, is given by

$$MC = 1.2(R + U + F) + 2.5L + 0.36\frac{I}{Q} \qquad (2.26)$$

where F is the container or packing cost per unit of product and Q is the annual production of units of the product. The constants 1.2, 2.5 and 0.36 are based on experience.

**Table 2.17: Components of Manufacturing Cost (Woods, 1975)**

| Direct Costs | Indirect Costs |
|---|---|
| a) **Costs which vary linearly with production rate** | Safety and protection |
| Raw materials | General plant overheads |
| Utilities | Payroll |
| Packing, containers | Restaurant |
| Shipping | Recreation |
| Royalties (if not a lump sum) | Storage facilities |
| b) **Costs varying with production but not linearly** | |
| Operating labour | |
| Direct supervision | |
| Maintenance | |
| Plant supplies | |
| Laboratory analyses | |
| c) **Costs that are independent of Production Rate** | |
| Rent | |
| Insurance | |
| Taxes | |
| Depreciation | |

## Illustrative Example 2.7

100,000 units of a chemical intermediate is to be produced, annually, in a small plant. The raw material costs were estimated to be N5 million, utility costs N0.4 million, while packing was expected to cost N0.15 million. Labour cost per unit is estimated at N25. If the total fixed investment is to be N45 million, estimate the manufacturing cost per unit of product using the general equation method.

## Answer

The general equation method for estimating manufacturing cost is given by

$$MC = 1.2(R + U + F) + 2.5L + 0.36\frac{I}{Q} \qquad from \quad (2.26)$$

$$R = \frac{N5,000,000}{100,000} = N50 \qquad (1)$$

$$U = \frac{N400,000}{100,000} = N4 \qquad (2)$$

$$F = \frac{N150,000}{100,000} = N1.50 \qquad (3)$$

Substituting these in equation (2.26) we get

$$MC = 1.2(N50 + N4 + N1.50) + 2.5 \, x \, N25 + 0.36 x \frac{45,000,000}{100,000}$$

$$= N291.10 \, per \, unit \, of \, product \quad Ans$$

### 2.5.3:  The Factor Method for Estimating Direct Manufacturing Costs (Peters & Timmerhaus, 1985)

This method is, also, known as the detailed general equation method. It accounts for more items of manufacturing cost than the ordinary general equation method. The manufacturing cost, MC, is given, in this case, by

$$MC = \sum factor \, x \, multiplying \, element \qquad (2.27)$$

Here, the multiplying element can be the total manufacturing cost or its components such as labour cost, raw material cost, utility cost etc. The factor is some fraction by which the multiplying element is to be scaled to obtain its contribution to the manufacturing cost.

Factors in common use are based on experience. Some factors are expressed as fractions of total product manufacturing cost while some are specific to particular components of manufacturing cost. The different types of factors are listed in Tables 2.18 and 2.19.

**Table 2.18: Factors for Estimating Manufacturing Costs (Peters & Timmerhaus, 1985)**

| Item | Factor | To be multiplied by |
|---|---|---|
| **Direct Production Cost** | 0.6 | Total Product Cost* |
| Raw materials | 0.1-0.5 | Total Product Cost |
| Operating labour | 0.1-0.2 | Total Product Cost |
| Direct supervisory & Clerical labour | 0.1-0.25 | Total Product Cost |
| Utilities | 0.1-0.20 | Total Product Cost |
| Maintenance & repairs | 0.02-0.1 | Fixed capital investment |
| Operating supplies | 0.005 - 0.01 | Fixed capital investment |
| Laboratory charges | 0.1-0.2 | Operating Labour |
| Patents & Royalties | 0 - 0.06 | Total Product Cost |
| **Fixed Charges** | 0.1-0.2 | Total Product Cost |
| Depreciation | 0.1 | Fixed capital investment for machinery |
| Building depreciation | 0.02 - 0.03 | Building Costs |
| Local taxes | 01 - 0.04 | Fixed Capital Investment |
| Insurance | 0.001-0.004 | Fixed Capital Investment |
| Rent | 0.08-0.12 | Value of Rented Land & Buildings |
| **Plant Overheads** | 0.05-0.15 | Total Product Cost |

*the total product cost is the sum of manufacturing plus general expenses

## 2.5.4: The Method of Varying Production Rates for Estimating Direct Manufacturing Costs (Woods, 1975)

In this method, the manufacturing cost is expressed as a general function of production rate such that it is the sum of three costs:- those that are fixed, those which vary, linearly, with production rate and those which vary, but not linearly, with production rate.

If Q is the number of units produced, $C_F$ the fixed manufacturing cost, $C_L$ the manufacturing cost which varies linearly with production rate and $C_R$ the manufacturing cost which varies, but not linearly, with production rate, it follows that

$$C_L = aQ \quad \text{and} \quad C_R = b + cQ^{d+1} \tag{2.28}$$

where $a$ is the cost per unit when cost varies linearly with production rate, $b$ is the incurred costs when cost varies, but not linearly, with production rate while $c$ is the cost per unit when costs vary, but not linearly, with production rate, and $d$ is a constant under these conditions. The total manufacturing cost, MC, is then

$$MC = C_F + C_L + C_R = C_F + aQ + b + cQ^{d+1}$$
$$= (C_F + b) + aQ + cQ^{d+1} \tag{2.29}$$

The unit manufacturing cost, UMC, is, thus, given by

$$UMC = \frac{MC}{Q} = a + \frac{(C_F + b)}{Q} \, cQ^d \tag{2.30}$$

Values of a, b, c, and d are obtained from information available for the particular project.

**Table 2.19: Factors for Estimating Manufacturing Costs (Peters & Timmerhaus, 1985)**

| Item | Factor | To be multiplied by |
|---|---|---|
| **A: Variable Production Costs** | | |
| Raw materials | 1 | Price per unit and number of units |
| Miscellaneous materials | 0.1 | Total variable costs |
| Utilities | 1 | Price per unit and number of units |
| Shipping & packing | 1 | Price per unit and number of units |
| **Sub-Total A** | | |
| **B: Fixed Costs** | **Factor** | **To be multiplied by** |
| Operating Labour | 1 | Operating labour |
| Supervision | 0.2 | Operating labour |
| Maintenance | 0.05 – 0.1 | Fixed Capital Investment |
| Plant Overheads | 0.05 | Fixed Capital Investment |
| Capital Charges | 0.15 | Fixed Capital Investment |
| Royalties | 0.01 | Fixed Capital Investment |
| Rates | 0.02 | Fixed Capital Investment |
| Insurance | 0.01 | Fixed Capital Investment |
| **Sub-Total B** | | |
| **Direct Production Cost** | **= A + B** | |

## 2.5.5: The Anergy Method for Estimating Direct Manufacturing Costs (Gaensslen, 1979)

The production cost can, also, be estimated using a refined anergy method, discussed earlier, for estimating capital investment. For a plant, manufacturing organic products,

$MC = fixed\ cos ts\ per\ gJ\ of\ end\ product$

$+ var\ iable\ cos ts\ per\ gJ\ of\ end\ product$

That is

$$. \ MC = Z.a\left(b + \frac{\alpha_P}{1 - \alpha_P}\right).\left(\frac{1 - \alpha_T . \alpha_P}{\alpha_T . \alpha_P}\right) + \frac{\beta}{\alpha_T . \alpha_P} \qquad (2.31)$$

where $Z$ is the capital load or capacity utilisation factor and $\beta$ is the cost of the organic raw material per $gJ$.

## 2.5.6: Other Methods of Estimating Direct Manufacturing Costs

All the methods, given so far, do not estimate other direct manufacturing costs such as taxes, rent and depreciation. The reason is that some of these, such as taxes and depreciation, can be estimated, directly, if the prevailing government policy in their regard, is known.

1.    <u>Taxes</u>

There are four basic types of tax namely:

- Property tax
- Excise and or value added tax (VAT)
- Income tax
- Social security tax

Property taxes are, usually, levied by local governments and may be different for residential, commercial and industrial property. Since property tax rates may vary with local government area, it is advisable to obtain the applicable rate for the area where the factory is to be located. Property tax is, usually, less than 10% of property value.

Excise tax is levied, in Nigeria, by the federal government. In some other advanced countries, state governments, also, levy excise tax. Excise tax is levied by governments on companies for the priviledge of letting the companies do business in their area of authority. Excise tax

can be levied on either manufactured or retailed goods. Prior to the
1990s, the federal government operated the manufacturer's excise tax
system. Since the late 1990s, however, the system in operation in
Nigeria is the excise tax on retailed goods or the value added tax, VAT.
The current rate is 5% on sales of manufactured goods and much higher
for some retailed items such as cigarettes, petrol and alcohol.

Income tax is the tax on gross profit. Gross profit is the difference
between total income or revenue and total manufacturing cost. Income
tax is the main source of revenue for governments which do not operate
a socialist system of governance. In Nigeria, only the federal
government can levy income tax on companies.

There are many forms of income tax.

*Normal tax* is the income tax levied on gross profit. It is, currently,
45%.

*Surtax* is an income tax on gross earnings which exceed a certain limit
imposed by law. The surtax rate may be higher or lower than the normal
income tax rate depending on the objective which the government
wishes to achieve through its tax policy.

*Capital gains tax* is the tax levied on profits made from the sale of
capital assets such as land, buildings or equipment. Distinction is,
usually, made between long term and short term capital gain so that
capital gains tax is calculated from a consideration of net long term
capital gain and net short term capital loss. Capital gains apply,
however, to companies, already, in operation.

2.      Insurance

Insurance is a cover or hedge taken either to protect against property
loss or charges based on legal liability. Often it is a legal requirement.
The major insurance requirements for manufacturing companies are
(Peters & Timmerhaus, 1985):

i.      Fire insurance on buildings, equipment and other property
ii.     Public liability (third party) insurance for bodily injury, property
        loss or damage
iii.    Business interruption insurance
iv.     Insurance to cover against special operating hazards such as occur

|       | in operating power plant and machinery |
| v.    | Workmen's compensation insurance |
| vi.   | Marine and transportation insurance |
| vii.  | Comprehensive crime insurance |
| viii. | Employee benefit insurance such as life, medical, accident and pension plan insurance. |

The cost of insurance is, usually, estimated at 1% of fixed capital investment.

## 3.    Depreciation

Depreciation is a means of spreading the initial capital expense on an asset over its useful life. The need for this arises from the fact that the asset depreciates in value over time. The method of estimating the depreciated value of an asset depends on how fast the value of that asset is expected to depreciate. Generally, the useful or depreciable life of an asset is specified by government policy. What is specified by government, however, may or may not coincide with the real life of the asset. Table 2.20 lists the government approved depreciable life for industries and assets in the USA and Nigeria. While USA practice makes provisions for depreciating assets in whole industries, Nigerian practice seems to allow only the depreciation of individual assets. Estimation of depreciation is important in the life of an industry because it allows the business to write off the cost of its investment in a manner approved by government.

**Depreciation Methods**

There are several methods for estimating depreciation in commercial organizations. Some of the methods are best suited to a business about to be started, some to businesses already in operation, while some are best suited to specific industries. The most commonly used depreciation methods in the chemical and allied industry are

|       |                                                      |
| i.    | The straight line method |
| ii.   | The declining balance or fixed percentage method |
| iii.  | The sum of the years digit method |
| iv.   | The sinking fund method |

## 3a. The Straight Line Method

This method allows the cost of the asset to be spread, uniformly, over its depreciable life. This is done by charging equal amounts of depreciation each year throughout the entire life of the asset. That is

$$Depreciation = \frac{Initial\ asset\ \cos t - Salvage\ value}{Depreciable\ life\ in\ years} = \frac{V_0 - V_N}{N} \quad (2.32)$$

Here $V_0$ is the initial asset cost, $V_N$ is the salvage value of the asset and N is the depreciable life in years.

**Table 2.20:    Depreciable Life for Industries and Individual Items in the USA and Nigeria (Perry & Green, 1984; Peters & Timmerhaus, 1985; Annual Reports of Nigerian Companies)**

| Industry (USA) | Depreciable life, years (USA) | Individual Items (Nigeria) | Depreciable Life, years (Nigeria) |
|---|---|---|---|
| Chemicals & allied products; Leather & leather products; Plastic products | 11 | Land; Buildings* | 50 |
| Fabricated metal products; Food products | 12 | Plant & Equipment; Furniture | 10 |
| Glass and glass products; Primary metal (non-ferrous); Rubber products | 14 | Vehicles | 4 |
| Stone & Clay products | 15 | | |
| Paper and allied products; Petroleum refining | 16 | | |
| Primary metal (ferrous); Sugar & sugar products; Vegetable oil products | 18 | | |
| Cement | 20 | | |

* If the lease has less than 50 years to expire, depreciation is over the period of the remaining lease. The method used depends on the accounting policies allowed by the government and adopted by the firm.

## 3b. The Declining Balance Method

This method seeks to account for the rapid decrease, in the early years, of the salvage value of the asset. A constant percentage of the remaining book value of the asset is written off each year. Then

$$Depreciation = k\left( \frac{Initial\ asset\ \cos t - Cummulative\ Depreciation}{Depreciable\ life\ in\ years} \right) \quad (2.33)$$

where $k$ is a constant, equal to 1 in the basic method, and to 2 in the double declining balance method. For example, if the initial asset cost was N100,000 with a depreciable life of 10 years. In the basic method, depreciation in the first year, $D_1$ is

$$D_1 = 1.0x \left( \frac{N100,000 - 0}{10} \right) = N10,000 \qquad (1)$$

In the second year, depreciation, $D_2$ is

$$D_2 = 1.0x \left( \frac{N100,000 - (N10,000 + 0)}{10} \right) = N9,000 \qquad (2)$$

In the third year, depreciation, $D_3$ is

$$D_3 = 1.0x \left( \frac{N100,000 - (N10,000 + N9,000 + 0)}{10} \right) = N8,100 \qquad (3)$$

Equations (1), (2) and (3) suggest that this calculation can be done according to a more general and useful form. Thus, if we denote the constant percentage factor, by which the initial value of the asset is depreciated, as $f$, at the end of the first year,

$$D_1 = f.V_0 \qquad (4)$$

The asset value at the end of the first year, $V_1$ is given by

$$V_1 = V_0 - D_1 = V_0 - f.V_0 = V_0(1 - f) \qquad (5)$$

At the end of the second year,

$$D_2 = f.V_1 = f.V_0(1 - f) \qquad (6)$$

With the asset value, $V_2$ given by

$$V_2 = V_0 - D_1 - D_2 = V_0 - f.V_0 - f.V_0(1 - f)$$
$$= V_0(1 - 2f + f^2) = V_0(1 - f)^2 \qquad (7)$$

Similarly, at the end of the third year,

$$D_3 = f.V_2 = f.V_0(1 - f)^2 \qquad (8)$$

The asset value, $V_3$ is given by

$$V_3 = V_0 - D_1 - D_2 - D_3 = V_0 - f.V_0 - f.V_0(1 - f) - f.V_0(1 - f)^2$$
$$= V_0(1 - 3f + 3f^2 - f^3) = V_0(1 - f)^3 \qquad (9)$$

Hence, generally, in any year, $i$, the depreciation, $D_i$, and the asset value, $Vi$, are given by

$$D_i = f.V_0(1 - f)^{i-1} \qquad (2.34)$$
$$V_i = V_0(1 - f)^i \qquad (2.35)$$

If N years represent the depreciable life of the asset, the salvage value at the end of this period, $V_S$, is given as

$$V_S = V_0(1 - f)^N \qquad from \quad (2.35)$$

53

from which

$$f = 1 - \left(\frac{V_S}{V_0}\right)^{\frac{1}{N}} \qquad (2.36)$$

### Illustrative Example 2.8

Calculate the salvage value and the depreciation, in the salvage year, of an asset with an initial value of N100,000, a depreciable life of 10 years and a fixed percentage factor of 10%. Use both the basic and the double declining balance methods.

### Answer

Using the basic declining balance method, from equation (2.35)

$$V_S = V_0 (1 - f)^N = N100,000 \times (1 - 0.1)^{10} = N34,867.84 \quad Ans.$$

From equation (2.34)

$$D_N = f.V_0 (1 - f)^{N-1} = 0.1 \times N100,000 \times (1 - 0.1)^{10-1} = N3874.20 \; Ans.$$

Using the double declining balance method, $f = 2 \times 0.1 = 0.2$, so that, as above,

$$V_S = V_0 (1 - f)^N = N100,000 \times (1 - 0.2)^{10} = N10,737.42 \quad Ans.$$

From equation (2.34)

$$D_N = f.V_0 (1 - f)^{N-1} = 0.2 \times N100,000 \times (1 - 0.2)^{10-1} = N2,684.36 \; Ans.$$

This example shows that the value of the asset is not depreciated to zero at the end of the depreciable life as is the case in the straight line method.

To depreciate the salvage value of an asset to zero, the combination method is, often, employed. In this method, the declining balance method is used for some part of the depreciable life while the rest of the depreciable life is handled by the straight line method.

### 3c. The Sum of the Years Digit (SYD) Method

This method gives results similar to the declining balance method in the sense that the value of an asset is, rapidly, depreciated in the early years of its depreciable life. The method is, also, able to depreciate the salvage value of the asset to zero or to any chosen value at the end of the depreciable life. It is, preferably, applied to assets with depreciable lives of about three years or so.

The annual depreciation rate, $f_i$, in any year, $i$, is determined from

$$f_i = \frac{number\ of\ useful\ depreciable\ years\ remaining}{sum\ of\ arithmetic\ series\ with\ depreciable\ life\ as\ limit} \quad (2.36)$$

For example, if the life of the asset, $N = 5$ years, the sum of the years is $1 + 2 + 3 + 4 + 5 = 15$.

As an arithmetic series, this sum is also $= \dfrac{n(n+1)}{2} = \dfrac{5 \times 6}{2} = 15$.

The value of $f$ in any year, $n$, in between, is

$$f_i = \frac{N - (n-1)}{N(N+1)/2} \quad (2.37)$$

Thus, in the first year, $n = 1$ and the depreciation rate is

$$f_1 = \frac{N - (n-1)}{N(N+1)/2} = \frac{5 - (1-1)}{5(5+1)/2} = \frac{5}{15}$$

In the second year, $n = 2$ and the depreciation rate is

$$f_2 = \frac{N - (n-1)}{N(N+1)/2} = \frac{5 - (2-1)}{5(5+1)/2} = \frac{4}{15}$$

and so on. An expression for the depreciation, $D_i$, in the sum of the years digit method, can then be given by

$$D_i = \frac{N - (n-1)}{N(N+1)/2} \times (V_0 - V_S) \quad (2.38)$$

## 3d. The Sinking Fund Method

This method involves the use of compound interest. It aims to accumulate sufficient funds so that the original investment capital can be recovered. This is done by setting aside a fixed amount of money every year (an annuity) such that at the end of the depreciable life, the sum of all monies set aside plus accrued interest will equal the amount of depreciation.

In any year, $i$, the amount of depreciation is equal to $V_0 - Vi$. If D is the amount of money set aside every year for N years, the first amount set aside, at the end of the first year, will earn interest for N-1 years. If the interest rate is $r$, then D will have accumulated to the value $D(1 + r)^{N-1}$ at the end of the depreciable life.

Similarly, the accumulated value of the second amount set aside in the second year, for N - 2 years, will be $D(1 + r)^{N-2}$, the third amount set aside in the third year, for N - 3 years, will be $D(1 + r)^{N-3}$ and so on.

The sum, S, of all accumulated amounts (the amount of the annuity) will be

$$S = D(1+r)^{N-1} + D(1+r)^{N-2} + D(1+r)^{N-3} \ldots + D(1+r) + D \quad (1)$$

Multiplying equation (1) by $(1 + r)$, we get

$$S(1+r) = D(1+r)^{N} + D(1+r)^{N-1} + D(1+r)^{N-2} \ldots + D(1+r)^{2}$$
$$+ D(1+r) \quad (2)$$

Subtracting equation (1) from equation (2) and re-arranging, we get

$$S = \frac{D\left[(1+r)^{N} - 1\right]}{r} \quad (2.39)$$

But $S = V_0 - V_S$. Hence

$$D = \frac{r(V_0 - V_S)}{(1+r)^{N} - 1} \quad (2.40)$$

Equation (2.39) also shows that the amount of depreciation, accumulating in any year, $i$, is equal to

$$S_i = V_0 - V_i = \frac{D\left[(1+r)^{i} - 1\right]}{r} \quad (2.41)$$

Substituting the value of D from equation (2.40) and re-arranging,

$$V_i = V_0 - (V_0 - V_S)\frac{(1+r)^{i} - 1}{(1+r)^{N} - 1} \quad (2.42)$$

### Illustrative Example 2.9

An equipment, bought for N200,000 is to be depreciated over a ten year period by means of the sinking fund method. If the salvage value is N8,000 determine the constant amount to be set aside for depreciation if the going interest rate is 25%. What is the asset value in the ninth year?

### Answer

Since $\quad D = \dfrac{r(V_0 - V_S)}{(1+r)^{N} - 1} \qquad$ from (2.40)

and $V_0 = $ N200,000; $V_S = $ N8,000; $r = 0.25$; $N = 10$ years, then

$$D = \frac{r(V_0 - V_S)}{(1+r)^{N} - 1} = \frac{0.25 \times (N200,000 - N8,000)}{(1+0.25)^{10} - 1} = N5,773.93 \ Ans.$$

The asset value in the 9th year is given, from equation (2.42) by

$$V_i = V_0 - (V_0 - V_S)\frac{(1+r)^i - 1}{(1+r)^N - 1}$$

$$= N200,000 - (N200,000 - N8,000)\frac{(1+0.25)^9 - 1}{(1+0.25)^{10} - 1} = N51,019.15 \; Ans.$$

Other Depreciation Methods

Other depreciation methods include the

1.      Units of Production Depreciation

Here the annual depreciation, D is

$$D = \frac{\cos t \; of \; fixed \; assets - residual \; value}{estimated \; total \; production} \; x \; actual \; production \quad (2.43)$$

2.      Units of Time Depreciation

For this

$$D = \frac{\cos t \; of \; fixed \; assets - residual \; value}{estimated \; completion \; time} \; x \; actual \; completion \; time \quad (2.44)$$

3.      Group Depreciation

This is useful in multiple asset accounting. The assets are, usually, similar in nature and have approximately the same useful lives. Straight line depreciation is used.

4.      Composite Depreciation

This is useful in multiple asset accounting. The assets are, usually, not similar in nature and have different useful lives. Straight line depreciation is used.

5.      Tax Depreciation

This is a tax deduction for recovery of the cost of assets used in the business or for the production of income. A common version of tax depreciation is the capital allowance by means of which a business can deduct a fixed percentage of the cost of depreciable assets every year.

## 2.6: Estimating General Expenses

As listed in Table 2.5, general expenses consist of administrative costs, distribution and selling costs, research and development costs as well as financing costs. In developing countries, very little attention tends to be paid to research and development so that, almost, no expense is made in this area. Also, only the minimum of investment is made in distribution and selling costs with customers, often, having to bear the cost of collecting the product from the factory site. In these cases, therefore, administrative costs tend to dominate the general expense portfolio.

In the more developed countries, however, administrative costs are about 2 - 6% of total product cost. Administrative cost, here, includes executive salaries, clerical wages, legal fees, office supplies and communication. Distribution and selling costs can be about 2 - 20% of total product cost and include costs for sales offices, sales men and women, shipping and advertising. Research and development costs are estimated at either 2 - 5% of sales or 5% of total product cost. Financing or interest on borrowed funds, though a fixed charge, is added to general expenses. It is estimated at between 0 and 10% of total capital investment. These figures are summarised in Table 2.21.

**Table 2.21: Estimates of Components of General Expenses**

| Item | % of total production cost | Item | % of total production cost |
|------|---------------------------|------|---------------------------|
| Administrative Expense | 2 - 6 | Research & Development | 5 or (2 – 5 % of sales) |
| Distribution and Selling Costs | 2 - 20 | Financing Costs | 0 - 10 (of total capital investment) |

## 2.7: Estimating Sales Income or Revenue

The estimation of income for a chemical manufacturing company or project can be both simple and complicated. The largest proportion of income for such a company or project comes, usually, from the sale of manufactured products. Such income is called sales income and is the product of selling price and the quantity of goods sold. That is

$$Income = Number\ of\ units\ sold \times Selling\ price\ per\ unit \quad (2.45)$$

Traditionally, the selling price is determined by the cost-plus method. That is

$$Selling\ price\ = manufacturing\ \cos t + profit + allowance\ for\ tax \quad (2.46)$$

In sophisticated markets, selling price is determined, in the market place, additionally, by such things as competition, demand, and value as perceived by the buyer.

The problem arises when the number of units sold, say, per annum, is to be, realistically, determined. For a product already in the market but for which the project is a new entrant or supplier in the market place, the orientation of the market study would be to determine who the competing producers are, their market share and their perceived marketing strategy *vis - a - vis* product pricing and technical or customer support services they offer. The determination of expected sales volume becomes one of deciding on a realistic but targeted market share and the means to achieve that share.

Another source of income, for a chemical company, is the sale of a service or services. Such a service may arise from the need to facilitate the use of the company's other products or of a particular expertise within the company. Service income is estimated either on some percentage basis, on a fixed charge for each service, on a cost-plus basis or on a period charge/retainership basis.

A chemical company can, also, earn income from financial investments. Chemical companies, often, find it expedient to invest in stock, debentures, etc., in other companies and projects, preferably but not necessarily, related to their operations in any manner. Such income, often, come, periodically, in the form of dividends, as some percentage of the amount invested.

Other income to the company can come from the sale of company assets. These assets could be land, buildings, equipment, equity holdings, or even, whole chemical plants. The asset value, at the time of sale, may have appreciated or been depreciated while cash receipts may be on lump sum or instalmental basis.

## 2.8: *Estimating the Profitability of Proposed Investments*

In analysing the profitability of a proposed investment, the basic issue that must be kept in mind is whether any alternative action other than the proposed investment would be more profitable. Profitability can take a narrow meaning as the excess of income over costs or the, increasingly, popular social meaning of the opportunity for creating jobs, positive environmental and economic impact, better military security, etc. The analysis here is restricted to the narrow meaning of financial profitability.

There are six common methods of assessing financial profitability; namely:

- Break - even Analysis (BEA)
- Return on Investment (ROI)
- Pay-out Time (POT)
- Net Present Worth (NPW) or Net Present Value (NPV)
- Discounted Cash Flow Rate of Return (DCFRR)
- Capitalised Costs (CC)

Each of these is treated, briefly, below. More detailed treatment may be found in the literature cited at the end of this chapter.

### 2.8.1: Break -Even Analysis

This is, perhaps, the most well known method. It is a worst-case type of analysis and is, entirely, deterministic. The purpose of the analysis is to find that quantity of production for which sales income equals total manufacturing costs. This is the break-even production. Production above this level will result in profit and below this level, in a loss. Figure 2.1 shows a typical break-even chart.

Mathematically, the break - even analysis may be done as follows. If $s$ is the selling price per unit of product and $Q$ the annual production of units, then, from equation (2.45)

$$Sales\ Income = Q \times s \qquad (2.45a)$$

From equation (2.29)

$$MC = (C_F + b) + aQ + cQ^{d+1} \qquad (2.29)$$

At break-even, sales income is equal to manufacturing costs, so that

$$Q_{break-even} \times s_{break-even} = (C_F + b) + aQ_{break-even} + cQ_{break-even}^{d+1}$$

$$Or\ cQ_{break-even}^{d+1} + Q_{break-even}(a - s_{break-even}) + (C_F + b) = 0 \qquad (2.47)$$

Equation (2.47) can be solved for the break-even production quantity or for the break-even sales price.

Profit, P, is, generally, given as the excess of sales income over costs. That is

$$P = Q \ x \ s - \left(C_F + b\right) - aQ - cQ^{d+1} \qquad (2.48)$$

Optimum profit is determined as, usual, from

$$\frac{d\,P}{d\,Q} = s - a - c\left(d+1\right)Q^d = 0 \qquad (2.49)$$

and

$$\frac{d^2\,P}{d\,Q^2} = -d\,c\left(d+1\right)Q^{d-1} = 0 \qquad (2.50)$$

Since the second derivative is negative, the Q obtained would be the optimum. Solving equation (2.49) we get, for a given sales price,

$$Q_{optimum} = \left(\frac{s-a}{c\left(d+1\right)}\right)^{\frac{1}{d}} \qquad (2.51)$$

or, for a given production quantity,

$$s_{optimum} = a + c\left(d+1\right)Q^d = 0 \qquad (2.52)$$

**Figure 2.1: A Break - Even Chart**

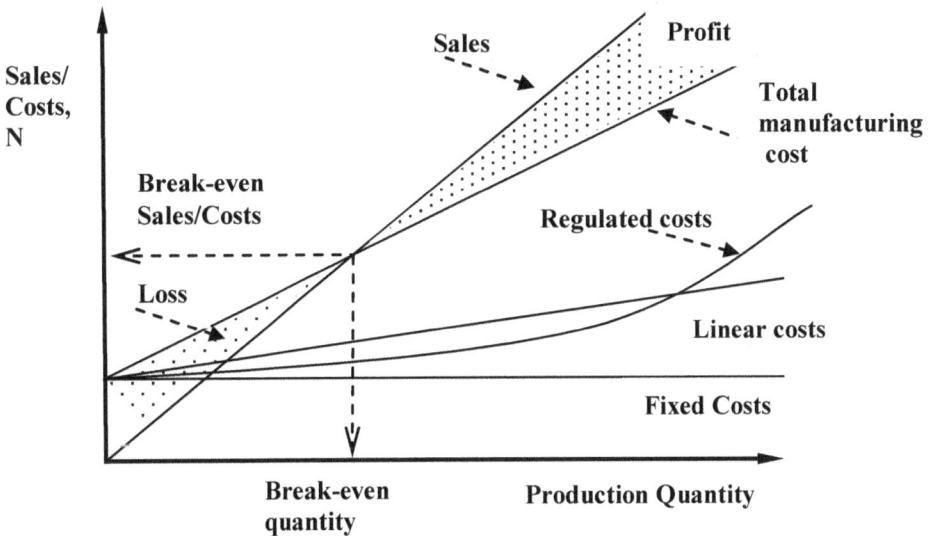

**Illustrative Example 2.10**

A plant is being planned to manufacture a product which sells for N100 per unit. The manufacturing costs per unit of product were estimated as

61

follows :-

- Fixed costs = N4.50
- Linearly varying costs = $N\,25\,Q$
- Other varying costs = $N\left(50 + 0.002\,Q^{1.9}\right)$

Q is the number of units manufactured per year. Determine the break-even production quantity and the production quantity for optimum profit.

**Answer**

At break-even, sales income = total manufacturing costs. That is

$$100\,Q_{break-even} = 4.5 + 25\,Q_{break-even} + 50 + 0.002\,Q_{break-even}^{1.9} \qquad (1)$$

This can be re-arranged to obtain

$$0.002\,Q_{break-even}^{1.9} - 75\,Q_{break-even} + 54.5 = 0 \qquad (2)$$

For ease of analysis, equation (2) may be taken as a quadratic by rounding 1.9 to 2. Then it will have the solution

$$Q_{break-even} = \frac{75 \pm \sqrt{75^2 - 4 \times 0.002 \times 54.5}}{2 \times 0.002} = 37,499 \quad units \qquad Ans.$$

The optimum profit is obtained when, from equation (2.51), the production quantity is

$$Q_{optimum} = \left(\frac{s-a}{c(d+1)}\right)^{\frac{1}{d}} = \left(\frac{100-25}{0.002(0.9+1)}\right)^{\frac{1}{0.9}} = 59,229 \; units \quad Ans.$$

### 2.8.2:  The Return on Investment (ROI) Method

In this method, the average yearly profit is calculated as a percentage of the capital invested. This gives some idea of the return on investment. That is

$$ROI = 100\,\frac{Average\ yearly\ profit\ in\ earning\ life}{Fixed\ investment + Working\ capital} \qquad (2.53)$$

In the Engineer's ROI method, the definition is slightly different. That is

$$ROI = 100\,\frac{Average\ yearly\ profit\ in\ earning\ life}{Original\ fixed\ investment + Working\ capital} \qquad (2.54)$$

This method is, also, known as the *operator*'s method and the ROI calculated as the *capitalised earning rate*. Because the original investment is involved, it is a measure of the return to be expected from an amount of money, initially, invested in the project.
In the Accountant's method,

$$ROI = 100 \frac{Average\ yearly\ profit\ in\ earning\ life}{Average\ (fixed\ investment\ +\ working\ capital)} \qquad (2.55)$$

The ROI calculated by this method is, also, known as the Return on Average Investment and the method as the Return on Book Investment method.

## Illustrative Example 2.11

In a certain project, the yearly investment in fixed and working capital, as well as the net profit plus depreciation, over a five year period, were estimated as shown in the table below. Determine the return on investment of this project using both the Engineer's and the Accountant's methods.

**Table of Investment and Profits for Project X**

| End        of | Fixed Capital, | Working |   | Net Profit + |
|---|---|---|---|---|
| Year | N | Capital, | N | Depreciation, N |
| 1 | 500,000 | 80,000 | | 75,000 |
| 2 | 100,000 | 50,000 | | 100,000 |
| 3 | | 70,000 | | 120,000 |
| 4 | 250,000 | 60,000 | | 130,000 |
| 5 | | 50,000 | | 140,000 |

## Answer

To calculate the return on investment using the Engineer's method, use equation (2.54) as follows:

Average yearly profit
$$= N \frac{75,000 + 100,000 + 120,000 + 130,000 + 140,000}{5} = N113,000 \qquad (1)$$
Original fixed investment + Working capital

$$= N500,000 + N80,000 = N580,000 \qquad (2)$$

Hence, from (2.54) and (1) and (2)

$$ROI = 100 \, x \frac{N113,000}{N580,000} = 19.48\% \qquad Ans$$

To use the Accountant's method, we calculate
Average yearly profit

$$= N \frac{75,000 + 100,000 + 120,000 + 130,000 + 140,000}{5} = N113,000 \quad (3)$$

Average (fixed investment + working capital)

$$= N1,000 \, x \frac{(500 + 80) + (100 + 50) + (0 + 70) + (250 + 60) + (0 + 50)}{5} = N232,000 \, (4)$$

Hence, from (2.55) and (3) and (4)

$$ROI = 100x \frac{N113,000}{N232,000} = 48.71\% \qquad Ans.$$

### 2.8.3: Pay-out Time Method

The payout time is the number of years required to recover the original depreciable investment from accruing profits, plus depreciation. This payout time is, also, referred to as the *pay-off period*, the *cash recovery period* or the *pay-back period*. Thus, in a chosen period,

$$Pay - out \; Time = \frac{Original \; depreciable \; fixed \; investment}{Average \; profit + Annual \; average \; depreciation} \qquad (2.56)$$

**Illustrative Example 2.12**

Calculate the pay-out time for the project of Example 2.11.

**Answer**

$$Average \; yearly \; profit = N \frac{75,000 + 100,000 + 120,000 + 130,000 + 140,000}{5}$$

$$= N113,000 \qquad (1)$$

$$Original \; depreciable \; fixed \; investment = N500,000 \qquad (2)$$

Note that working capital is not a depreciable investment since depreciation is not allowed on working capital. Hence, from (2.56) and (1) and (2)

$$Pay - out \; Time = \frac{N500,000}{N113,000} = 4.42 \; years \quad Ans.$$

64

## 2.8.3.1: Pay-out Time Method with Interest

Because money has a time related value, pay-out times, calculated without considering the time value of money, may be unrealistic. Thus, it is supposed that a more realistic time may be obtained if the original investment is allowed to attract interest at a rate equivalent to a minimum acceptable rate of return. In addition, the average profit per year and the average depreciation per year are regarded as annuities. Then

*Payout time with interest*

$$= \frac{depreciable\ fixed\ capital\ investment + interest}{average\ (profit + annual\ depreciation)as\ an\ annuity} \qquad (2.57$$

### Illustrative Example 2.13

Calculate the pay-out time, with interest, for the project of Example 2. 11. Assume an interest rate of 19% for the fixed capital and for the annuity. The service life of the project is 5 years.

### Answer

Let $n$ be the service life = 5
At the end of this period, the depreciable fixed investment, FC, will be worth

$$FC\,(1+r)^n = N500{,}000\,(1+0.19)^5 = N1{,}193{,}176.83 \qquad (1)$$

Let R be the annuity, or the annual constant income from profits and depreciation, which will compound to the total amount of profits, $S_i$, at the rate of 19 % at the end of the year, $n$. The total amount of profits, $S_i$, in any year, $i$, is given by

$$S_i = \frac{R\left[(1+r)^i - 1\right]}{r} \qquad (2)$$

derived from the sinking fund method of estimating depreciation, equation (2.39), by replacing D by R and ($V_0$ - $V_i$) by $S_i$. When $i = 5$,

$$S_i = N(75{,}000 + 100{,}000 + 120{,}000 + 130{,}000 + 140{,}000) = N565{,}000 \quad (3)$$

Since r = 0.19,

$$R = \frac{r\,S_i}{(1+r)^i - 1} = \frac{0.19\,x\,N565{,}000}{(1+0.19)^5 - 1} = N77{,}433.34\ per\ annum \qquad (3)$$

Hence, from (2.57), (1) and (3)

$$Pay-out\ time\ with\ int\,erest = \frac{N1{,}193{,}176.83}{N77{,}433.34} = 15.41\,years\ \ Ans.$$

## 2.8.4:  The Present Value Method

The present value of an investment is a consequence of the time value of money. That is, a Naira in hand today is worth more than a Naira in six month's time because of its prospective yield or appreciation (not: by inflation) if it was invested in some profitable activity. For example, the present value of N100 received in future, at different rates of interest, are as follows as shown in Table 2.22.

**Table 2.22: Table of the Present Value of N100 at Various Rates of Interest**

| Beginning of year on which N100 was received | Rate of Interest (Return) | | | |
|---|---|---|---|---|
| | 0% | 5% | 10% | 20% |
| | P resent | | V alue, | N |
| 1 | 100 | 100 | 100 | 100 |
| 2 | 100 | 95.25 | 90.91 | 83.33 |
| 3 | 100 | 90.70 | 82.64 | 69.44 |
| 5 | 100 | 82.27 | 68.30 | 48.23 |

In other words, to get N100 in two year's time, you would have to invest, now, N100 at 0 % interest or N90.70 at 5 % interest or N82.64 at 10 % interest and so on. These values were obtained by multiplying the principal, N100, by $1/(1+r)^n$ where $r$ is the interest rate (in decimals) and $n$ is the number of years. The product of principal and interest is., also, known as the discounted value. That is

$$Discounted \ \ value = \frac{Pr incipal}{(1+r)^n} \qquad (2.58)$$

$1/(1+r)^n$ is known as the discount or present value factor and standard tables list its numerical values against $n$ and $r$.

The net present value (NPV) is, then, the sum of the annual discounted values over the number of years, $n$. That is

$$NPV = \sum_i^n Discounted \ \ value \ in \ year \ i \qquad (2.59)$$

## 2.8.5:  The Discounted Cash Flow Rate of Return Method

The objective, in this method, is to search for that interest rate, $r$, which equalises the present worth of cash inflows (profit, depreciation, etc) and cash outflows (investments, etc) such that the sum of all NPVs is zero. That is

$$\frac{A_0}{(1+r)^0} + \frac{A_1}{(1+r)^1} + \frac{A_2}{(1+r)^2} + \ldots \ldots \frac{A_N}{(1+r)^N} = 0 \qquad (2.60)$$

This rate is the discounted cash flow rate of return (DCFRR). $A_0$, $A_1$, $A_2$, $A_N$ are the annual cash flows in the years 0, 1, 2.....N.

The decision to invest will, then, be influenced by whether this interest rate of return is less than, equal to or greater than some minimum acceptable or hurdle rate of return. In low risk environments, 15 % is the hurdle rate (the median in the Fortune 500, in 1978, was, in fact, 14.3 % (Fortune Magazine, July 1979)). In moderately risky environments, the rate is about 25 % while in high risk environments, it can be 50 % or more per year.

### Illustrative Example 2.14

Calculate the DCFRR for a project for which an initial cash outlay of N100,000 yields an anticipated cash inflow, over four years, as follows

| Year | 2 | 3 | 4 | 5 | Total |
|------|------|------|------|------|------|
| Cash inflow | N45,000 | N40,000 | N35,000 | N30,000 | N150,000 |

### Answer

The first step in solving this problem is to select factors, from a table of present value factors, for two rates of interest between which the anticipated rate may be expected to lie. In this case, let us choose 10 % and 20 % interest rates. The table below is, then, constructed.

| Year | Cash flow in , current year, N | PV factor @ 10 % | PV of cash flow @ 10%, N | PV factor @ 20% | PV of cash flow @ 20%, N |
|------|------|------|------|------|------|
| 1 | -100,000 | 1.0000 | -100,000 | 1.0000 | -100,000 |
| 2 | +45,000 | 0.9091 | +40,909.50 | 0.8333 | 37,498.50 |
| 3 | +40,000 | 0.8264 | +33,056 | 0.6944 | 27,776 |

| | | | | | |
|---|---|---|---|---|---|
| 4 | +35,000 | 0.7513 | +26,295.50 | 0.5787 | 20,254.50 |
| 5 | +30,000 | 0.6830 | +20,490 | 0.4823 | 14,469 |
| Net cash flow | 50,000 | | 20,751 | | -2 |

Since the sum of all PV at 20 % interest rate is -2 (acceptably, close to zero for amounts in thousands), the desired interest rate of return (CFRR) is 20 %. Ans.

Note that in some cases, the answer may not come out so easily as in the above example. Such cases, often, require interpolation or trial and error procedures.

### 2.8.5.1: General Methods of Calculating Cash Flow Rate of Return

The previous example was used to illustrate the basic principles and procedures for calculating the discounted cash flow rate of return. In practice, more complicated data on inflows and outflows are, usually, encountered making it necessary to develop more structured and systematic procedures for handling such problems.

The basic elements in the calculation are the cash inflows and cash outflows. Cash inflows include profits before taxes plus depreciation while cash outflows consist of manufacturing or operating costs, investment less depreciation plus working capital. A typical calculation algorithm, for doing a cash flow rate of return, is set out in Table 2.23 below.

Most modern computer spreadsheet software include both public domain and specialist algorithms for the computation of cash flow rate of return. In some companies and organisations, standardised sheets are used in line with the peculiar objectives and methods of operation of the organization.

**Table 2.23: Typical Calculation Algorithm for the Cash Flow Rate of Return (CFRR)**

| S/No | Quantity to be calculated | S/No. | Quantity to be calculated |
|---|---|---|---|
| 1 | Sum of operating expenses = Outflow | 8 | Cash inflow after Taxes = item 3 minus item 7 |
| 2 | Capital outlay + sum of | 9 | Net cash outflow = item 2 |

|   | | | | |
|---|---|---|---|---|
|   | expenses = Total outflow | | | minus item 8 |
| 3 | Cash inflow before taxes | 10 | | Net cash inflow = item 8 minus item 2 |
| 4 | Depreciation | 11 | | Choose a rate of return, $r$, in decimal units |
| 5 | Total cost = Item 1 plus item 4 | 12 | | Set Sum = $(1 + r)$ x item 9 minus item 10 |
| 6 | Tax base = Item 3 minus item 5 | 13 | | Repeat from item 11 until Sum is zero |
| 7 | Tax = 0.55 x item 6 | | | |

## Illustrative Example 2.15

Determine the cash flow rate of return for a project whose expenditure and revenue pattern is shown below.

| Timing | Expenditures, N | Receipts, N |
|---|---|---|
| Start 2000 | 100,000 | |
| Start 2001 | 150,000 | |
| End 2001 | | 90,000 |
| End 2004 | | 150,000 |
| End 2005 | | 250,000 |
| Totals | 250,000 | 490,000 |

## Answer

To solve this problem, a reference point is, first, selected, most, advisedly, the beginning of the first year of receipts. The effect of the reference point is to influence the difficulty of the calculations not the correctness of the answer obtained. Thus

| Timing | Expenditures, N | Receipts, N | PV factors @ 30 % interest |
|---|---|---|---|
| Start 2000 | 100,000 | | 1.10 |
| Start 2001 | 150,000 | | 1.00 |
| End 2001 | | 90,000 | 0.91 |
| End 2004 | | 150,000 | 0.68 |
| End 2005 | | 250,000 | 0.62 |
| Totals | 250,000 | 490,000 | |

A project evaluation sheet, that enables trial calculations to be made, can be constructed as shown in Table 2.24 below.

### Table 2.24: Project Evaluation Sheet for Example 2.15

| Timing | Trial 1 @ 0 % interest | Trial 2 | @ 10% interest | Trial 3 | @ 20% interest | Trial 4 | @ 30% interest |
|---|---|---|---|---|---|---|---|
| Period | Actual disburse-ment, N | Factor | Present Worth, N | Factor | Present Worth, N | Factor | Present Worth, N |
| Before 1st. Year | 100,000 | 1.10 | 110,000 | 1.20 | 120,000 | 1.30 | 130,000 |
| After 1st. Year | 150,000 | 1.00 | 150,000 | 1.00 | 150,000 | 1.00 | 150,000 |
| Total (A) | 250,000 | | 260,000 | | 270,000 | | 280,000 |
| Period | Actual Receipts, N | Factor | Present Worth, N | Factor | Present Worth, N | Factor | Present Worth, N |
| End of 1st Year | 90,000 | 0.909 | 81,810 | 0.833 | 74,970 | 0.769 | 69,210 |
| End of 2nd Year | - | 0.826 | - | 0.694 | - | 0.592 | . |
| End of 3rd Year | - | 0.751 | - | 0.579 | - | 0.455 | . |
| End of 4sh Year | 150,000 | 0.683 | 102,450 | 0.482 | 72,300 | 0.350 | 52,500 |
| End of 5th Year | - | 0.621 | - | 0.402 | . | 0.269 | . |
| End of 6th Year | 250,000 | 0.565 | 155,250 | 0.335 | 100,500 | 0.207 | 67,250 |
| End of 7th Year | - | 0.513 | - | 0.279 | - | 0.159 | - |
| End of 8ih Year | - | 0.467 | - | 0.233 | - | 0.123 | . |
| End of 9th Year | - | 0.424 | - | 0.194 | - | 0.094 | - |

| End of 10th Year | - | 0.386 | - | 0.162 | - | 0.073 | - |
|---|---|---|---|---|---|---|---|
| **Total (B)** | 490,000 | | 339,510 | | 247,770 | | 188,960 |
| **Ratio (A/B)** | 0.51 | | 0.77 | | 1.093 | | 1.49 |

If the DCFRR is plotted against the ratio, A/B, as shown in the Figure below, it is found that the required DCFRR for A/B to be unity is 17%.

Note, also, that this method takes into account the following:

   a.  Expenditures are made at more than one time
   b.  Amount of expenditure may vary
   c.  Receipt's timing can have gaps
   d.  Amount of receipts may, also, vary

The basic assumptions are that

 i.  All net cash outflows must precede receipts
 ii.  Mutually exclusive alternatives must be appraised by evaluation of the difference between alternatives
 iii.  All financing costs must be excluded from the evaluation since the interest rate of return may be viewed as some cost of capital

**Plot of DCFRR against the Ratio of Cash Outflow and Cash Inflow**

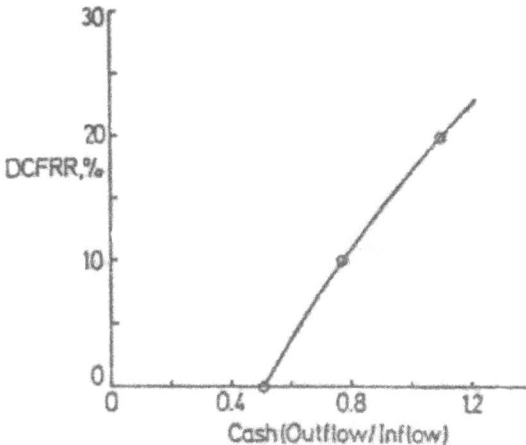

71

## Illustrative Example 2.16

A labour saving device is proposed which can be developed for N15.000, built for N165,000 and which will result in income, from reduction in the cost of labour and related employee benefits, of N1.00 per unit of output. It will take one year to develop and build this equipment. In view of the technical developments in product and process, a useful depreciable life of 5 years is assumed. The project revenue and expenditure, based on anticipated output, are as follows:

|        | Development Expense | Construction Expense | Total Outlay | Income (Cost Reduction), N |
|--------|---------------------|----------------------|--------------|----------------------------|
| 1991   | N15,000             | N165,000             | N180,000     | -                          |
| 1992   |                     |                      |              | 90,000                     |
| 1993   |                     |                      |              | 100,000                    |
| 1994   |                     |                      |              | 110,000                    |
| 1995   |                     |                      |              | 110,000                    |
| 1996   |                     |                      |              | 110,000                    |

Determine the cash flow rate of return based on incremental expenditure and income for this project.

## Answer

The project evaluation sheet is prepared as shown in the Table below. The balance outstanding to next year at 20% is -N8543, which has crossed zero indicating an overshoot of the correct DCFRR. Several trials, finally, give the balance outstanding to next year as N36, indicating that the required DCFRR is 10.7 % since N36 is as good as zero in consideration of the thousands involved.

Sometimes, it is desired to determine the sales price or capacity utilisation which will result in a certain DCFRR. This type of problem is illustrated in Example 2.17 below.

### Project Evaluation Sheet for Example 2.16

|   |                     | 1991 N | 1992 N | 1993 N | 1994 N | 1995 N |
|---|---------------------|--------|--------|--------|--------|--------|
|   | **Outflow**         |        |        |        |        |        |
| A | Development Expense | 15,000 |        |        |        |        |
| B | Capital             |        |        |        |        |        |

| | | | | | | |
|---|---|---|---|---|---|---|
| | Expense | | | | | |
| C | Total Outflow | 165,000 | | | | |
| D | Cash Inflow before Taxes | 180,000 | 90,000 | 100,000 | 110,000 | 110,000 |
| | **Calculation of Tax** | N | N | N | N | N |
| E | Expense Outlay | 15,000 | | | | |
| F | SYD Depreciation | | 55,000 | 44,000 | 33,000 | 22,000 |
| G | Total Cost (E+F) | 15,000 | 55,000 | 44,000 | 33,000 | 22,000 |
| H | Basis for Tax, (D-G) | (15,000) | 35,000 | 56,000 | 77,000 | 88,000 |
| I | Tax @ 0.55H | - | 19,250 | 30,800 | 42,350 | 48,400 |
| J | Inflow after Tax (D-I) | - | 70,750 | 69,200 | 67,650 | 61,600 |
| | **Net Cash Flow after Tax** | N | N | N | N | N |
| K | Outflow(C-J) | 171,750 | | | | |
| L | Inflow (J-C) | - | 70,750 | 69,200 | 67,650 | 61,600 |
| | **Trial 1** | | | | | |
| | Balance forward from prior year | | 171,750 | 135,350 | 93,220 | 44,214 |
| | plus interest @ 20% | | 34,350 | 27,070 | 18,644 | 8,843 |
| | less cash inflow balance outstanding to next year | | 70,750 | 69,200 | 67,650 | 61,600 |
| | | 171,750 | 135,350 | 93,220 | 44,214 | -8,543 |
| | **Trial 2** | | | | | |
| | Balance forward from prior year | | 171,750 | 126,763 | 76,577 | 20,414 |
| | plus interest @ 15% | | 25,763 | 19,014 | 11,487 | 3,062 |
| | less cash inflow balance outstanding to next year | | 70,750 | 69,200 | 67,650 | 61,600 |
| | | 171,750 | 126,763 | 76,577 | 20,414 | -55,476 |

73

| Trial 3 | | | | | |
|---|---|---|---|---|---|
| Balance forward from prior year | | 171,750 | 118,175 | 60,793 | |
| plus interest @ 10% | | 17,175 | 11,818 | 6,079 | |
| less cash inflow balance outstanding to next year | | 70,750 | 69,200 | 67,650 | |
| | 171,750 | 118,175 | 60,793 | -778 | |
| **Trial 4** | | | | | |
| Balance forward from prior year | | 171,750 | 119,377 | 62,950 | |
| plus interest @ 10.7% | | 18,377 | 12,773 | 6,736 | |
| less cash inflow balance outstanding to next year | | 70,750 | 69,200 | 67,650 | |
| | 171,750 | 119,377 | 62,950 | 36 | |

## Illustrative Example 2.17

An investment of N55,000 is to be made over 2 years in a project for which an 7% investment credit is made. It is assumed that working capital requirements amount to 20% of sales, operating costs to 40k per unit of product. The estimated useful life of the project is 10 years. Market forecasts indicate that the following sales volume forecasts may be used. Determine the sales price to give a 20% DCFRR.

### Sales Volume Forecasts for Example 2.17

| Year | 75% Capacity Utilisation | 85% Capacity Utilisation | 100 % Capacity Utilisation |
|---|---|---|---|
| | Sales in Operating Year, | Thousands of Units | |
| 1 | 8 | 10 | 40 |
| 2 | 12 | 25 | 40 |
| 3 | 15 | 35 | 45 |
| 4 | 25 | 40 | 50 |

| 5 | 35 | 40 | 50 |
| 6 | 40 | 45 | 50 |
| 7 | 45 | 50 | 50 |
| 8 | 50 | 50 | 50 |
| 9 | 50 | 50 | 50 |
| 10 | 50 | 50 | 50 |

## Answer

First, the calculation is performed for 100% capacity utilisation as shown in the project evaluation sheet below.

**Project Evaluation Sheet for Example 2.17, Part A, 100% Capacity**

| | A | B | C | D | E | F |
|---|---|---|---|---|---|---|
| Year | Investment, | Investment Credit, N | SYD, N | Working Capital, N | Units Sold | Revenue, (ExX), N |
| 0 | .30,000 | 2,100 | | -8,000X | | |
| 1 | -25,000 | 1,750 | | | | |
| 2 | | | 10,000 | | 40,000 | 40,000X |
| 3 | | | 9.000 | | 40,000 | 40,000X |
| 4 | | | 8,000 | | 45,000 | 45,000X |
| 5 | | | 7,000 | | 50,000 | 50,000X |
| 6 | | | 6,000 | | 50,000 | 50,000X |
| 7 | | | 5,000 | | 50,000 | 50,000X |
| 8 | | | 4,000 | | 50,000 | 50,000X |
| 9 | | | 3,000 | | 50,000 | 50,000X |
| 10 | | | 2,000 | | 50,000 | 50,000X |
| 11 | | | 1,000 | | 50,000 | 50,000X |
| Termi nal Value | | | | 10,000X | | |
| | 55,000 | 3,850 | 55,000 | 2,000X | 475,000 | 475,000X |

Next, the calculation is performed for 100% capacity utilization with the desired present value (PV) factor of 20 % as shown in the project evaluation sheet below.

75

## Project Evaluation Sheet for Example 2.17, Part B, 100% Capacity, PV Factor = 20%

| | G | H | I | J | K | |
|---|---|---|---|---|---|---|
| Year | Operating costs, (0.4E), N | Profit before tax, (F-G), N | Profit after tax, (0.45H), N | Undiscounted cash flow, (A+B+C+D+I), N | PV factor @ 20% | Discounted cash flow, (JxK), N |
| 0 | | | | -27,900 | 1.000 | -27,900 |
| 1 | | | | -23,250 | 0.833 | 19,367 |
| 2 | 16,000 | 40,000X– 16,000 | 18,000X -7,200 | 10,000X +2,800 | 0.674 | 6,940X + 1943 |
| 3 | 16,000 | 40,000X– 16,000 | 18,000X -7,200 | 18,000X +1,800 | 0.579 | 10,422X + 1042 |
| 4 | 18,000 | 45,000X– 18,000 | 20,250X - 8,100 | 20,250X - 100 | 0.482 | 9,761X -48 |
| 5 | 20,000 | 50,000X - 20,000 | 22,500X -9,000 | 22,500X -2,000 | 0.402 | 9,045X-804 |
| 6 | 20,000 | 50,000X - 20,000 | 22,500X- 9,000 | 22,500X -3,000 | 0.335 | 7,538X - 1,005 |
| 7 | 20,000 | 50,000X - 20,000 | 22,500X - 9,000 | 22,500X - 4,000 | 0.279 | 6,278X - 1,116 |
| 8 | 20,000 | 50,000X - 20,000 | 22,500X - 9,000 | 22,500X - 5,000 | 0.233 | 5,243X- 1,165 |
| 9 | 20,000 | 50,000X - 20,000 | 22,500X - 9,000 | 22,500X - 6,000 | 0.194 | 4,365X- 1,164 |
| 10 | 20,000 | 50,000X - 20,000 | 22,5- 9,00000X | 22,500X - 7,000 | 0.162 | 3645X - 1,134 |
| 11 | 20,000 | 50,000X - 20,000 | 22,500X - 9,000 | 22,500X - 8,000 | 0.135 | 3038X - 1,080 |
| Terminal value | | | | 10,000X | 0.135 | 1,350X |
| | 190,000 | 475,000X - 190,000 | 215,750X - 81,650 | 215,750X - 81,650 | | 67,625X - 51,798 |

The sales price, SP, is then obtained from $67,625 \, x \, X - 51,798 = 0$ from which $\quad X = \dfrac{N\,51{,}798}{67{,}625} = N0.77 = 77\,kobo \quad Ans.$

## 2.8.6: Method of Capitalised Costs

The capitalised cost of an equipment is the theoretical total amount of the initial capital required to purchase the equipment plus that required to replace it at the end of its service life. If perpetuity is defined as an annuity in which periodic payments continue indefinitely, the capitalised cost can, also, be defined as the original cost of the equipment plus the present value of the renewable perpetuity (Peters & Timmerhaus, 1985).

Let PV be the present worth which will accumulate to an amount, S, during $n$ interest periods, at a periodic interest rate, $r$. Then

$$S = PV\,(1+r)^n \qquad (2.61)$$

If $C_R$ is the replacement value of the equipment, and if perpetuation is to occur,

$$PV = S - C_R \qquad (2.62)$$

In other words, the difference between, S, the amount accumulated at any time, $n$, and $C_R$, the replacement value, must equal the present worth, PV. Eliminating S from equations (2.61) and (2.62), we get

$$PV = S - C_R = PV\,(1+r)^n - C_R$$

or $\qquad\qquad PV = \dfrac{C_R}{(1+r)^n - 1.} \qquad (2.63)$

The capitalised cost, $C_C$, is then given, if $C_0$ is the original cost of the equipment, by

$$C_C = C_0 + \dfrac{C_R}{(1+r)^n - 1.} \qquad (2.64)$$

## Illustrative Example 2.18

An electric motor was bought for N5.000. Its replacement value, at the end of 10 years, was put at N6,500. If the interest rate is 15%, compounded every year, determine the capitalised cost of this motor.

## Answer

From equation (2.64),

$$C_C = C_0 + \frac{C_R}{(1+r)^n - 1.} = N5,000 + \frac{N6,500}{(1+0.15)^{10} - 1.} = N7,134.26 \quad Ans.$$

The capitalised cost method is most useful in comparing investment alternatives. The investment which results in the least capitalised cost is to be preferred. Various relationships can be derived between capitalised cost, $C_C$, replacement value, $C_R$ and the scrap value, $V_S$, of an equipment One of these relationships, using the capitalised cost factor, $f_C$, is derived below.

Equation (2.64) can be re-arranged as follows:

$$C_C = C_0 + \frac{C_R}{(1+r)^n - 1.} = \frac{C_0\left((1+r)^n - 1\right) + C_R}{(1+r)^n - 1} \tag{2.65}$$

If the scrap value, Vs, is equal to the difference between original and replacement costs (this is not always the case), then

$$V_S = C_0 - C_R \quad or \quad C_0 = V_S + C_R \tag{2.66}$$

Substituting (2.66) in (2.65),

$$C_C = \frac{(V_S + C_R)\left((1+r)^n - 1\right) + C_R}{(1+r)^n - 1} = \frac{V_S\left((1+r)^n - 1\right) + C_R(1+r)^n}{(1+r)^n - 1}$$

$$= V_S + \frac{C_R(1+r)^n}{(1+r)^n - 1} = V_S + f_C C_R \tag{2.67}$$

where $f_C$ is the capitalised cost factor, given by

$$f_C = \frac{(1+r)^n}{(1+r)^n - 1} \tag{2.68}$$

When operating costs are to be included in the investment analysis, each annual operating cost is treated as equivalent to a piece of equipment that lasts one year. In such a case, the analysis is as follows:

Let $C_N$ be the annual operating expense. The present value of this annual cash expense is

$$PV = \sum_{i=1}^{N} \frac{C_N}{(1+r)^N} \tag{2.69}$$

The capitalised present value is

$$Capitalised\ PV = f_C . PV = \left(\frac{(1+r)^N}{(1+r)^N - 1}\right) . \sum_{i=1}^{N} \frac{C_N}{(1+r)^N} \tag{2.70}$$

In simple cases, where $C_N$ is constant,

$$PV = \sum_{i=1}^{N} \frac{C_N}{(1+r)^N} = \frac{C_N}{r} \qquad (2.71)$$

so that the total capitalised cost, which includes operating cost, becomes

$$Capitalised \ cost = V_S + \left( \frac{C_R (1+r)^N}{(1+r)^N - 1} \right) + \frac{C_N}{r} + working \ capital \qquad (2.72)$$

## Illustrative Example 2.19

Determine the total capitalised cost of an investment which is to be made as follows.

- total fixed capital is N5million,
- working capital, N250,000, and
- annual operating expenses, N400,000

The service life of the project is estimated at 5 years, at which time the salvage value of the equipment will be N350,000. Interest rates are expected to be steady at 19% during this period.

## Answer

The replacement value, $C_R$, of the plant is assumed to be, from equation (2.66),

$$C_R = C_0 - V_S = N5,000,000 - N350,000 = N4,650,000 \qquad (1)$$

Hence, the capitalised cost of equipment, over 5 years, is given by

$$C_{C,equip} = V_S + \frac{C_R (1+r)^n}{(1+r)^n - 1} = N350,000 + \frac{N4,650,000 \times (1.19)^5}{(1.19)^5 - 1}$$

$$= N350,000 + N8,004,122.50 = N8,354,122.50 \qquad (2)$$

The capitalised cost of the operating expenses is given by equation (2.71) as

$$C_{C,op \, exp} = \frac{C_N}{r} = \frac{N400,000}{0.19} = N2,105,263.16 \qquad (3)$$

Since the annual working capital is N250,000, the total capitalised cost is given by equation (2.72) as

$$C_{C,total} = N8,354,122.50 + N2,105,263.16 + N250,000 = N10,709,385.66 \quad Ans.$$

## 2.9: Effect of Inflation

The foregoing calculations were made on the basis that there was no inflation and that only market rates of interest were considered. Yet inflation, whether general (open) or differential (repressed), can have telling effects on the profitability of investments. The effect of differential inflation is, generally, more pronounced than that of general inflation. All types of inflation, however, make a project appear more profitable than it, actually, is.

In general or open inflation, all costs and prices increase at a uniform rate. The calculated rate of inflation will be the same regardless of the particular mix of goods and services chosen. In differential or repressed inflation, the rate of inflation will depend on the spending pattern of the individual or company. This is the more common occurrence as a given company's labour and material costs may inflate at different rates. Government policy may result in repressed inflation as a result of taxation, import control, price controls, etc.

Although inflation rates and interest rates, in the market place, are related, they have different effects on the real value of future earnings. The effect of the market interest rate is that N1.00 invested today can be recouped next year as the sum of the Naira invested plus interest earned on it. The effect of inflation is that N1.00 earned next year, by whatever means, will buy less than N1.00 this year.

### 2.9.1: Effect of Inflation on Net Present Value (NPV)

The effect of inflation on net present value (NPV) can best be illustrated by looking at how it is calculated. Let $A_i$ be the cash flow in any given year, $i$, and $A_{DCFi}$, the discounted cash flow in the same year.

When there is no inflation, we have

$$NPV = \sum_{0}^{N} A_{DCFi} = A_0 + \frac{A_1}{(1+r)^1} + \frac{A_2}{(1+r)^2} + \ldots\ldots \frac{A_N}{(1+r)^N}$$

$$= A_0 + \sum_{1}^{N} \frac{A_i}{(1+r)^i} \tag{2.73}$$

where $r$ is the discrete, annual, compound, interest rate, that is, the cost of capital. It is, also, assumed that all interest payments occur at the end of the year.

When there is inflation at a rate, $r_f$, equation (2.73) becomes

$$NPV = A_0 + \sum_{1}^{N} \frac{A_i}{(1+r)^i (1+r_f)^y} \qquad (2.74)$$

## Example 2.20

Determine the present value, in Example 2.14, when the market rate of interest is 10% and the inflation rate 20%.

## Answer

The NPV table is constructed as follows

| Year | Cash Flow, N | Interest Rate Factor @ 10% | PV@10% Interest, N | Inflation Factor @ 20% | PV @ 20% Inflation, N |
|------|--------------|----------------------------|--------------------|------------------------|-----------------------|
| 0 | - 100,000 | 1.00 | -100,000 | 1.00 | -100,000 |
| 1 | + 45,000 | 0.91 | 40,950 | 0.83 | 33,989 |
| 2 | + 40,000 | 0.83 | 33,200 | 0.69 | 22,908 |
| 3 | + 35,000 | 0.75 | 26,250 | 0.58 | 15,225 |
| 4 | +30,000 | 0.68 | 20,400 | 0.48 | 9,792 |
|  | **150,000** |  | **20,800** |  | **-18,086** |

These figures show the present value in any year. When these are plotted, as in Figure 2.1 below, with the no inflation figures, the negative effects of inflation can be, more clearly, seen.

**Fig. 2.1: Effect of Inflation on Present Value of an Investment**

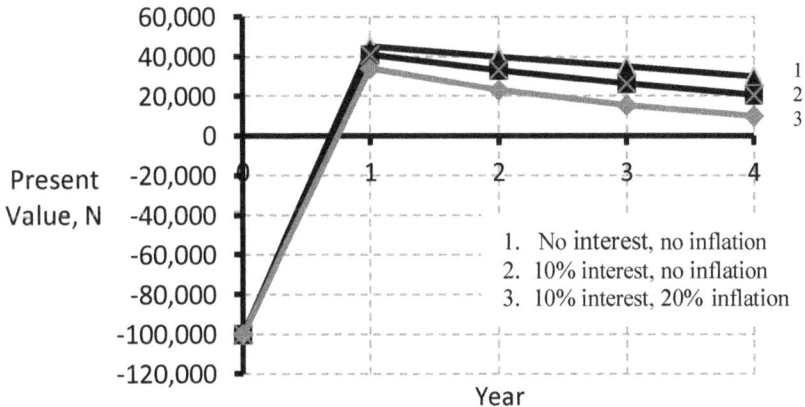

1. No interest, no inflation
2. 10% interest, no inflation
3. 10% interest, 20% inflation

### 2.9.2: Effect of Inflation on DCFRR

Since the DCFRR is that interest rate, $r_C$, which makes the NPV zero, we get, from equation (2.73), that

$$NPV = 0 = A_0 + \sum_{1}^{N} \frac{A_i}{(1+r)^i} \qquad (2.75)$$

At the end of any one year, a net annual cash value, $A_i$, will have a cash value of $A_i(1+r)$ at an interest rate of $r$. If the inflation rate is $r_f$, an effective rate of return, $r_{eff}$, can be defined such that

$$A_i\left(1 + r_{eff}\right) = \frac{A_i\left(1+r\right)}{\left(1+r_f\right)}$$

Rearranging, knowing that $r$ = DCFRR,

$$r_{eff} = \frac{\left(1+r\right)}{\left(1+r_f\right)} - 1 = \frac{1+r-1-r_f}{1+r_f} = \frac{r-r_f}{1+r_f} = \frac{DCFRR - r_f}{1+r_f} \qquad (2.76)$$

For example, if the DCFRR is 20 % and the inflation rate is 10%, then the effective or true DCFRR is, from (2.76)

$$r_{eff} = \frac{DCFRR - r_f}{1+r_f} = \frac{0.2 - 0.1}{1+0.1} = 0.09 \quad or \quad 9\% \qquad Ans$$

### 2.9.3: Effect of Inflation on Payback Period

As in the previous sections, when there is no inflation,

$$NPV = A_0 + \sum_{1}^{N} \frac{A_i}{(1+r)^i} = A_0 + \sum_{1}^{N} \frac{A_i}{(1+DCFRR)^i} \qquad from \quad (2.73)$$

The payback period, PBP, when there is no inflation, is

$$PBP = \frac{NPV - A_i}{A_i} = \sum_{1}^{N} \frac{1}{(1+DCFRR)^i} \qquad (2.77)$$

When there is inflation, from (2.76)

$$DCFRR = \left(1 + r_{eff}\right).\left(1 + r_f\right) - 1 \qquad (2.78)$$

### 2.10: Project Risk Analysis

The economic evaluation of a project cannot be complete without a project risk analysis. Every enterprise carries an element of risk. In project risk analysis, the extent of predetermined types of risk, usually financial, is ascertained. This enables the investor to make decisions concerning the types and levels of risk he or she is willing to take.

A summary of the more common methods of project risk analysis is given below. No attempt will be made to elaborate on the technical details of each method. These are, quite appropriately, dealt with in books dedicated to the subject.

## 2.10.1: Sensitivity Analysis

This identifies those variables for which variations or uncertainty in their estimation can have significant effects on the financial health of the project. With these variables and for the levels of variation or uncertainty, the economic evaluation is re-done, one at a time, or simultaneously if desired or possible. In the end, the analysis indicates where accurate estimates are vital. The major disadvantage of this method is that it deals with only the deterministic effects of variations or uncertainties in estimates but not with the chances of their occurrence.

## 2.10.2: The Hurdle Rate Analysis

In this method, a cut-off or hurdle rate (or level) in the DCFRR or NPV is set by company policy, usually, based on experience. For example, a DCFRR of 15 % for low risk, 25 % for moderately risky and 50 % for high risk environments, is often employed as cut-off levels. In other words, the riskier the environment, the more it must return.

## 2.10.3: Pay-off Table Analysis

Here, a project expectation is defined as the product of the NPV and its probability of occurrence, A pay-off table is then constructed for the situations in which there is competition and no competition which enables a decision to be more, easily, made.

## 2.10.4: The Mini-Max Loss Analysis

This method enables a course of action to be selected which minimises the maximum loss which could be sustained, even though this maximum loss may be a remote possibility. The analysis is, often, employed in a situation where the project would absorb a major part of the company's resources and project failure would result in bankruptcy of the company.

### 2.10.5: Utility Function Analysis

This analysis is done on the basis that the subjective value of money depends on the quantum of the amount to be invested in comparison to the amount available. For example, a loss of N10 to a man whose total wealth is N100 is not the same as the same loss to a man who has N100,000.

### 2.10.6: Probability Analysis

The premise, here, is that an estimate or forecast is, in fact, part of a probability distribution which expresses, subjectively, the relative chances that the variables of interest will have the estimated values. The profitability indices (NPV, DCFRR), calculated from these estimates, would, then, be obtained as probability distributions as well. The method is not practical except for very simple cases although with increasing sophistication of modem computers and mathematical methods, the situation is, gradually, changing.

### 2.10.7: The Monte Carlo Simulation

This is a more general method of handling stochastic data. Basically, the method selects values from a given probability distribution of project cash flows and uses these values to generate complete, yearly, net cash flows and cash flow rates or return (CFRR), also, as probability distributions. The method of selecting values from the probability distribution is by the use of random numbers and a cumulative distribution so that the more frequent estimates are selected more frequently for the analysis. The robustness of the final distribution, say, of the CFRR, is indicated when more iterations or change in sequence of random numbers do not affect the attribution.

## *References For Chapter Two*

1. Fortune Magazine (1979); *Fortune 's Directory of the 500 Largest Industrial Corporations*, Vol. 99, pp.265.295, Time Inc., NY, USA

2. Gaensslen H, (1979): *A Short Cut to Investment Costs*; Chemtech; May; pp 305 - 309; A.C.S.,Washington D.C, USA

4. Green D. W. (Editor); *Perry's Chemical Engineer's Handbook; 6th Edition*; Chapter 25; McGraw-Hill Book Co.; NY, USA (1984)

5. Haselbarth (1967) in Woods D. R.; *Financial Decision Making in the Process Industry*; Chapter 6; Prentice Hall Int'l; NJ, USA (1975)

6. Lyn & Howland (1960) in Woods D. R.; *Financial Decision Making in the Process Industry*: Chapter 6; Prentice Hall Int'l; NJ, USA ( 1975)

7. Mascio N. E. (1979); *Predict Costs Reliably via Regression Analysis*: Chemical Engineering; Feb12, Vol. 86 No. 4, pp 115 - 121; McGraw-Hill Book Co.; NY, USA.

8. Peters M. S, Timmerhaus K. D. ; *Plant Design and Economics for Chemical Engineers*; 3rd Edition; Chapters 4 to 9; McGraw - Hill Int'l Book Co,; London, UK (1985)

9. Pikulik A., Diaz H. E.,(1977); *Cost Estimating for Major Process Equipment*; Chemical Engineering; Oct. 10, Vol. 84 No. 21, pp 106-122; McGraw-Hill Book Co.; NY, USA

10. Sinnott R, K.; *Chemical Engineering, Vol. 6 (SI Units)* by J. M. Coulson and J. F, Richardson; An Introduction to Chemical Engineering Design; Chapter 6; Pergamon Press; Oxford, UK, (1983).

11. Woods D, R.; *Financial Decision Making in the Process Industry*; Chapter 6; Prentice Hall Int'l; NJ, USA (1975)

# CHAPTER THREE
# PROCESS DEVELOPMENT AND DESIGN

Once the idea of a new chemical product or process is conceived and taken seriously, the first sensible thing to do is **product** or **process research**. This research may be in the form of literature surveys and/or laboratory work. It provides information on material and system properties, methods of synthesis, purification, etc., of the product and conditions for carrying out the process.

Process research is then followed by an evaluation of the results of the research, called **research evaluation**. Information and data obtained are reviewed in terms of the economics, the unit operations and processes involved and all other information required for the design of the plant.

Through **process development**, the identified process steps are organized in a meaningful way to make it possible for the material and energy balances, as well as the material properties in the system, to be obtained. Often, the characteristics of the process, identified to be key, critical or unknown, are tested on a laboratory scale for small scale plants or in a **pilot plant** for large scale plants. The data obtained in such tests can be used as is for small scale plants or for scaling up the design for the large scale plant. The emergence of computer simulation, however, has reduced the use of pilot plant studies somewhat.

Through **process design**, the data obtained from process research and development are used to develop an efficient combination of sequences of the various processing steps in order to present a coherent view of the proposed chemical processing plant. This leads to an accurate determination of material and energy balances in the system.

Through **equipment design**, all the material and energy balances and material properties obtained in the above steps are translated into three dimensions as physical equipment with which each processing operation is carried out physically.

In **plant design**, all process equipment, associated pumps, heat exchangers, measuring and control instruments, etc, are arranged in the most viable configuration in which the chemical plant can be expected to operate as planned.

Small scale plants do not have the mixing non idealities of large scale plants. Thus, laboratory data on kinetics, thermodynamics and transport behaviour may be used straight away without much error. For large scale plants, however, scale up and similarity procedures must be employed.

The criteria, for determining when a process plant is small or large scale, is often based on government policy on paid up capital of operating companies or more realistically, from a chemical engineering point of view, on whether scale up techniques are necessary for the calculated performance of the process to match its actual operation.

This chapter assumes that process research, research evaluation and process development, including the associated pilot plant studies, have been done. Using the information obtained from these, an overview of process design steps and concepts is presented.

### 3.1: Process Design

The aim of process design is to arrange the processing sequences, identified for the particular process, in an optimal manner. It is assumed that the following information have been obtained in earlier studies (Vibrandt & Dryden, 1959):

| | |
|---|---|
| a) | A written description of the process |
| b) | Knowledge of all raw materials, products and intermediates |
| c) | Knowledge of physical characteristics and chemical compositions of all raw materials, products, and intermediates under the operating conditions of the process |
| d) | Knowledge of all process temperatures, pressures and concentrations |
| e) | Safety precautions, chemical reactions and properties of materials of construction |
| f) | Material and energy balances around each and every process or operation |
| g) | A complete diagramatic flowsheet showing the flows of process streams and the temperature, pressure at appropriate points. |

### 3.2: Process Design Steps

Four basic steps are necessary for a successful process design:

    i.       The first step is to prepare a flow sheet or a pictorial/diagramatic

           representation of the physical and chemical operations of the
process,

ii.       The second step is to carry out the material and energy balances
of the process using well known procedures,

iii.      The third step is to specify the equipment required in the process.
Estimates of the operating capacity of the equipment are made at
this stage,

iv.      The final step is to estimate the cost and profitability of the plant.

Flow sheet preparation, or flow sheeting, has meant different things to
different people. To some, flow-sheeting involves items (i), (ii) and
sometimes (iii) above, while for others, it is (i) and (ii) while for some
it is, simply, (i) above. The essential thing, however, is to reach step (iv)
in the most economical and efficient manner.

A similar procedure is described by Rudd et al (1978) in their book,
Process Synthesis. According to this method, the flow sheet is built up
after the following steps:

a)   reaction path synthesis,
b)   material flow synthesis,
c)   separation task selection
d)   task integration.

Systematic application of engineering judgement in specie allocation,
material balances, separation technology and engineering of process
systems result in a process design which is optimal.

This can be seen as another way of doing the same thing, this time, in a
task oriented, rather than in a procedure oriented manner. The task
oriented and the procedure oriented approaches have, each, their
advantages and disadvantages which depend on their suitability to
specific chemical process designs. Task oriented approaches would
appear to be more suitable for less complicated plants or to sections of
very large or complicated chemical plants.

### 3.2.1: Flowsheets

The flowsheet is a diagram which shows the arrangement of the
individual units of equipment selected to carry out the process, how
these units of equipment are connected to each other by streams of
material or energy which flow between them, and the flow rates,
composition and operating conditions of temperature and pressure of

those streams (Sinnott, 1983).

The flowsheet is, hence, a very important document in process design because it presents, at a glance, the critical information necessary for doing the following:

- Design of the piping, instrumentation and layout of the equipment and the plant
- Preparing the operating manuals for operating personnel or for operator training
- Comparing the actual operating performance of the plant with its expected or design performance, either during start-up of the plant or during its subsequent, normal operation.

### 3.2.2: Information required in all Flowsheets

For flowsheets to be useful, two kinds of information should, of necessity, be provided during their preparation. These are (Sinnott, 1983):-

a)    Essential information

    i)    Equipment layout. Here an equipment flowsheet provides complete information

    ii)    Stream composition, either as the flowrate of each component of the stream or as a mass, mole or volume fraction

    iii)    Total stream flow rate in kg/s, kmol/s., etc

    iv)    Nominal operating pressure, in bars, at every point where the pressure is different.

b)    Optional Information

These enhance the clarity of presentation and the usefulness of the flowsheet. Such optional information consist of things like:-

    i)    Physical property data such as density, viscosity, etc., as mean values for the streams

    ii)    Stream names

    iii)    Stream enthalpies

### 3.2.3: Types of Flowsheets

Many types of flowsheets are in use, depending on the stage of process design, on the use to which they are to be put and on the amount of information available for their preparation.

Block Diagram Flowsheets

At the initial stages of process design, when only the type and sequence of operations are known, the most useful and practical flowsheet is the block diagram form shown in Fig. 3.1a. This diagram illustrates, using rectangular or circular symbols, the sequence and types of operations of the process. Each block or circle represents a single piece of equipment or a complete stage in the process.

In the example given of a garri manufacturing process, the block diagram illustrates the sequence of cassava storage, washing, peeling, grating, pressing, fermentation, frying, sieving and packaging. The block diagram flowsheet is, especially, useful in making preliminary decisions in respect of sequencing of processing operations.

**Fig. 3.1a:  Block Diagram Flowsheet of Garri Manufacturing Process**

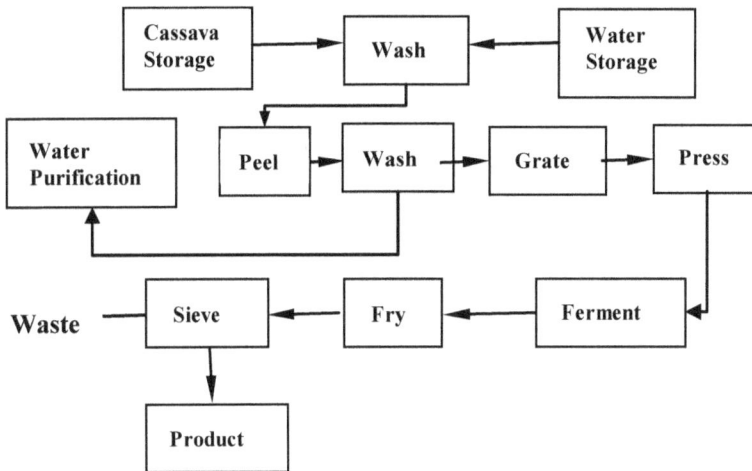

Fig. 3.1b shows another type of block diagram flowsheet and Fig. 3.1c yet another form of Fig. 3.1b. Either one of these types could be used in advertising the process. Each has the advantage of simplicity while outlining all the important operations of the process.

Equipment Flowsheet

This flowsheet presents the information in standardised form, using standard chemical engineering symbols. Each equipment, for a given

91

operation, for example, an evaporator, is drawn in the flowsheet using its standard symbol (Austin, 1979).

Further, each key operation is coded with a letter while the equipment for the operation is numbered. Evaporation can be coded, for example, with the letter, E, while the three evaporators used would be numbered E-1, E-2, E-3. Figure 3.2 shows a version of an equipment flowsheet.

Equipment flowsheets are, often, useful in discussions with management within the company since they look professional without being too complicated for management to understand the key issues. They are final enough, in terms of operation type and sequencing, to enable material balance calculations to be, accurately, made.

**Fig. 3.1b:  Another Type of the Block Diagram Flowsheet: The Synthesis of Ammonia by Steam Reforming Process (U.N.I.D.O., 1967)**

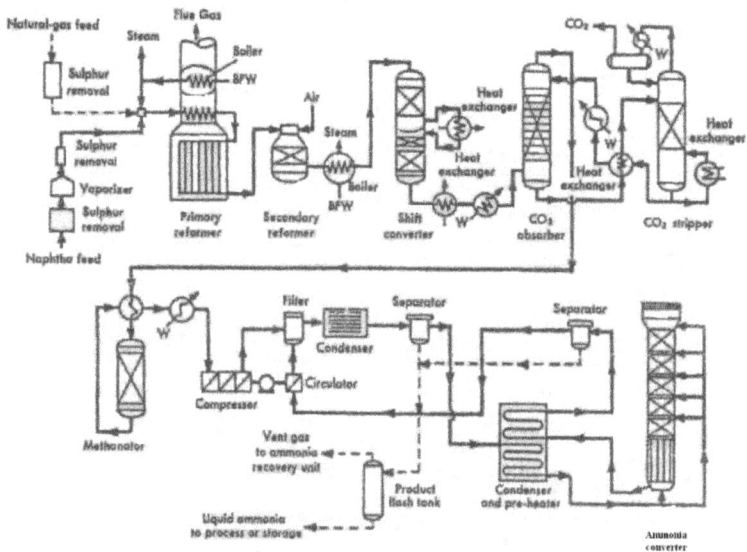

Material Balance Flowsheet

This flowsheet associates the equipment flowsheet with the material balances of the process in one diagram. Because of the variety and complexity of chemical engineering processes, several forms of the material balance flowsheet are in use, depending on company methods of operation and policy on the management and control of such information. However, three types of material balance flowsheets can be identified:-

92

i) *Quantitative flowsheet*. This gives a summary of the material balance calculations in tabular or block diagram form(see Fig. 3.3).

ii) *Quantitative graphic flowsheet*. This presents the same information (of the quantitative flowsheet) in an equipment flowsheet (see Fig. 3.4). All stream components, flow rates and compositions are placed at their proper locations in the diagram. Material balances are, also, included. These forms of the material balance flowsheet cannot be very convenient except for very simple processes. They are useful, however, within the process design group, in checking the accuracy of calculations and in rationalising process sequences in the light of new understanding provided by calculated results.

iii) *Complete material balance flowsheet*. This is the final flowsheet. It gives all the necessary information on the material flow in the process. It is presented in, fairly, standardised format, with equipment or processing step represented by standard symbols. Each equipment is coded with an alphabet and a number such as E-1, while each stream is numbered or its temperature/pressure listed. Four things are contained in the diagram, namely; the process flow diagram (equipment flowsheet), the key code (which lists what the coded letters represent), the table of material flow (which is sometimes omitted, and the proprietor/client/designer identity chart. Fig. 3.5 shows a complete material balance flowsheet containing all but the table of material flow.

The Energy Balance Flowsheet

This is, usually, developed after the material balance flowsheet has been determined. It gives the energy values associated with the streams, operations and individual equipment as well as the temperatures at these positions. The energy balance flowsheet can be presented in either flowsheet form as in Fig. 3.6a or in pictorial form as in Fig 3.6b, depending on the traditions of the industry. The modern trend is to list energy values with the material balance in tabular form, as in Table 3.1, and process and stream temperatures in a process flow diagram as in Fig. 3.4.

The Piping Flowsheet

This shows the piping connections between units together with the associated valves, machinery and other controls. They are, especially, useful in developing the eventual plant layout and cost of piping for the process. Again, depending on the complexity of the plant, the piping flowsheet may be done as in Fig. 3.7 or as part of the instrument and control diagram as in Fig. 3.8.

**Fig. 3.1c:** **Another Type of the Block Diagram Flowsheet: The Manufacture of Isoprene from Propylene (Scientific Design Co., 1968)**

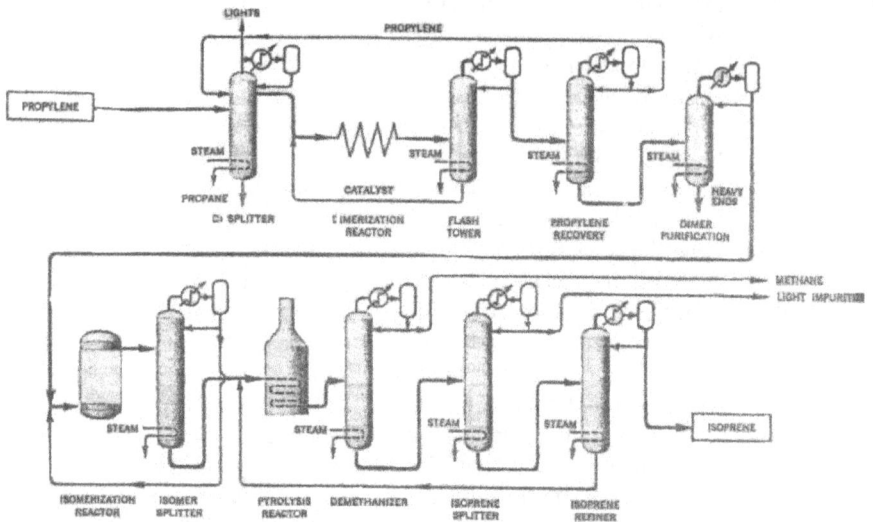

**Fig. 3.2: Equipment Flowsheet for Continuous Low Temperature Rendering System (Filstrup, 1976)**

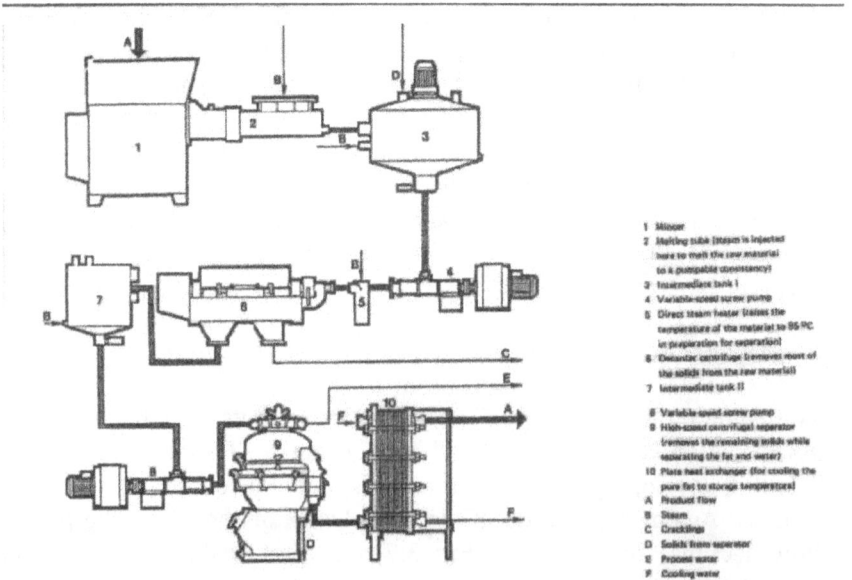

94

## Fig. 3.3: An Example of a Quantitative Flowsheet
## (Vilbrandt & Dryden, 1959)

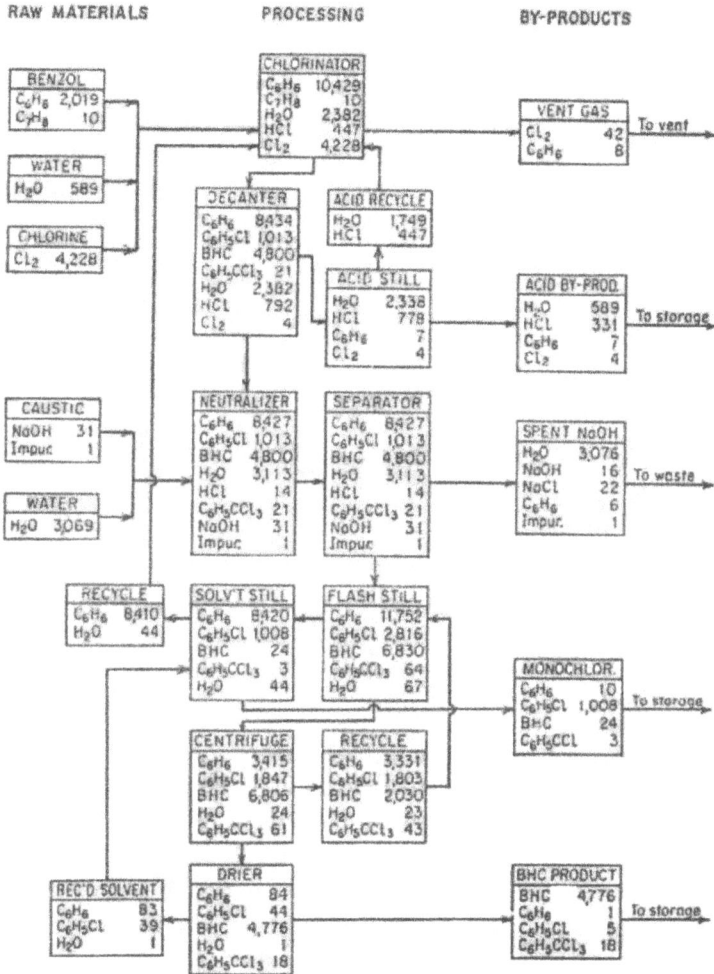

RAW MATERIALS      PROCESSING      BY-PRODUCTS

BENZOL
$C_6H_6$ 2,019
$C_7H_8$ 10

WATER
$H_2O$ 589

CHLORINE
$Cl_2$ 4,228

CHLORINATOR
$C_6H_6$ 10,429
$C_7H_8$ 10
$H_2O$ 2,382
HCl 447
$Cl_2$ 4,228

VENT GAS
$Cl_2$ 42
$C_6H_6$ 8
To vent

DECANTER
$C_6H_6$ 8,434
$C_6H_5Cl$ 1,013
BHC 4,800
$C_6H_4CCl_3$ 21
$H_2O$ 2,382
HCl 792
$Cl_2$ 4

ACID RECYCLE
$H_2O$ 1,749
HCl 447

ACID STILL
$H_2O$ 2,338
HCl 778
$C_6H_6$ 7
$Cl_2$ 4

ACID BY-PROD.
$H_2O$ 589
HCl 331
$C_6H_6$ 7
$Cl_2$ 4
To storage

CAUSTIC
NaOH 31
Impur 1

WATER
$H_2O$ 3,069

NEUTRALIZER
$C_6H_6$ 8,427
$C_6H_5Cl$ 1,013
BHC 4,800
$H_2O$ 3,113
HCl 14
$C_6H_5CCl_3$ 21
NaOH 31
Impur 1

SEPARATOR
$C_6H_6$ 8,427
$C_6H_5Cl$ 1,013
BHC 4,800
$H_2O$ 3,113
HCl 14
$C_6H_5CCl_3$ 21
NaOH 31
Impur 1

SPENT NaOH
$H_2O$ 3,076
NaOH 16
NaCl 22
$C_6H_6$ 6
Impur. 1
To waste

RECYCLE
$C_6H_6$ 8,410
$H_2O$ 44

SOLV'T STILL
$C_6H_6$ 8,420
$C_6H_5Cl$ 1,008
BHC 24
$C_6H_5CCl_3$ 3
$H_2O$ 44

FLASH STILL
$C_6H_6$ 11,752
$C_6H_5Cl$ 2,816
BHC 6,830
$C_6H_5CCl_3$ 64
$H_2O$ 67

MONOCHLOR.
$C_6H_6$ 10
$C_6H_5Cl$ 1,008
BHC 24
$C_6H_5CCl$ 3
To storage

CENTRIFUGE
$C_6H_6$ 3,415
$C_6H_5Cl$ 1,847
BHC 6,806
$H_2O$ 24
$C_6H_5CCl_3$ 61

RECYCLE
$C_6H_6$ 3,331
$C_6H_5Cl$ 1,803
BHC 2,030
$H_2O$ 23
$C_6H_5CCl_3$ 43

REC'D SOLVENT
$C_6H_6$ 83
$C_6H_5Cl$ 39
$H_2O$ 1

DRIER
$C_6H_6$ 84
$C_6H_5Cl$ 44
BHC 4,776
$H_2O$ 1
$C_6H_5CCl_3$ 18

BHC PRODUCT
BHC 4,776
$C_6H_6$ 1
$C_6H_5Cl$ 5
$C_6H_5CCl_3$ 18
To storage

## Fig. 3.4: An Example of a Quantitative Graphic Flowsheet
## (Vilbrandt & Dryden, 1959)

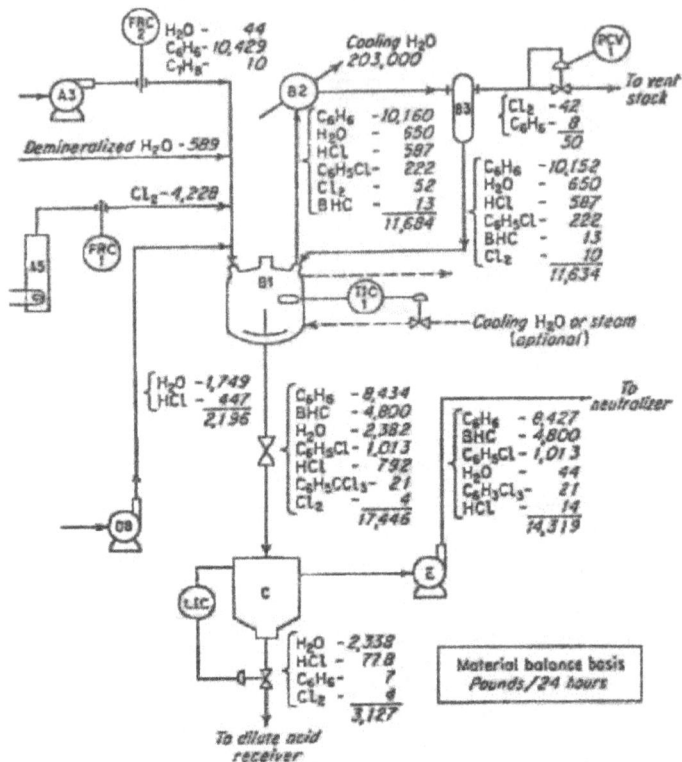

$H_2O$ - 44
$C_6H_6$ - 10,429
$C_7H_8$ - 10

Cooling $H_2O$ 203,000

To vent stack

Demineralized $H_2O$ - 589

$Cl_2$ - 4,228

$C_6H_6$ - 10,160
$H_2O$ - 950
$HCl$ - 587
$C_6H_5Cl$ - 222
$Cl_2$ - 52
$BHC$ - 13
11,684

$Cl_2$ - 42
$C_6H_6$ - 8
50

$C_6H_6$ - 10,152
$H_2O$ - 650
$HCl$ - 587
$C_6H_5Cl$ - 222
$BHC$ - 13
$Cl_2$ - 10
11,634

Cooling $H_2O$ or steam (optional)

$H_2O$ - 1,749
$HCl$ - 447
2,196

$C_6H_6$ - 8,434
$BHC$ - 4,800
$H_2O$ - 2,382
$C_6H_5Cl$ - 1,013
$HCl$ - 792
$C_6H_5CCl_3$ - 21
$Cl_2$ - 4
17,446

$C_6H_6$ - 8,427
$BHC$ - 4,800
$C_6H_5Cl$ - 1,013
$H_2O$ - 44
$C_6H_5Cl_3$ - 21
$HCl$ - 14
14,319

To neutralizer

$H_2O$ - 2,338
$HCl$ - 778
$C_6H_6$ - 7
$Cl_2$ - 4
3,127

To dilute acid receiver

Material balance basis
Pounds/24 hours

## Instrumentation and Control Diagram

Process control implies, usually, the control of three variables:- pressure, temperature and flow. Such control is achieved by the use of control instruments. Depending on the complexity of the process, the instrumentation and control diagram may or may not be incorporated into the piping flowsheet. Pressure and flow control can be, in some cases, entirely, incorporated into the piping diagram while the temperature control can be incorporated into the energy balance flowsheet. In more complex plants, greater clarity is achieved through separate flowsheets for each. Figure 3.8 illustrates a typical instrumentation and control diagram.

**Fig. 3.5:** **A Complete Material Balance Flowsheet (Example – the Synthesis of Ammonia by the Steam Reforming Process (U.N.I.D.O., 1967)**

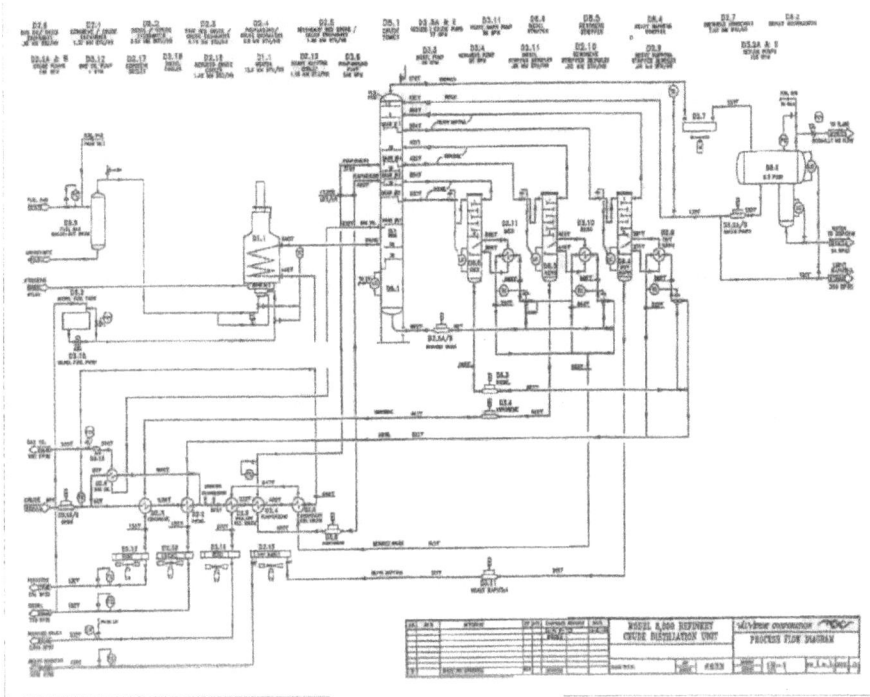

**Table 3.1: Heat, Material and Pressure Balance - Crude Distillation Unit, 5000 BPD-37.1 API (Val Verde Corp., 1999)**

| Description | o F | 0A.P. I. | Sp. Gr. 60 0F | UOP , K | Mol. Wt | BPSD @ 600F | GPM @ T&P | Lbs/ hr | Visco-sity @ T&P |
|---|---|---|---|---|---|---|---|---|---|
| Crude charge | 90 | 37.06 | 0.839 | 11.5 | 204.56 | 5000 | 151 | 61173 | 1.900 |
| Overhead Condensate | 130 | 91.71 | 0.634 | 13.3 | 87.09 | 3300 | 101.4 | 30456 | 0.395 |
| Atmos. tower o/head | 175.5 | 92.63 | | 13.38 | 85.45 | 3641.2 | | 33494 | |
| Atmos. tower reflux | 130 | 91.71 | 0.634 | 13.35 | 87.09 | 3202.4 | 98.5 | 29578 | 0.395 |
| L.P. Light naphtha | 130 | 91.71 | 0.634 | 13.35 | 87.09 | 95.1 | 2.9 | 878 | 0.395 |
| Reflux accum. | 130 | 101.86 | | 13.58 | 71.79 | 34.37 | | 3038 | |

97

| vapour | | | | | | | | | |
|---|---|---|---|---|---|---|---|---|---|
| H.P. Light naphtha | 130 | 100.93 | 0.609 | 13.55 | 72.69 | 327.7 | 10.2 | 2908 | 0.421 |
| Fuel gas | 130 | 122.51 | | 14.10 | 55.69 | 16.0 | | 130 | |
| Stripped heavy naphtha | 281 | 58.33 | 0.745 | 12.08 | 120.34 | 1135.6 | 38.47 | 12338 | 0.469 |
| Stripped kerosene | 465 | 41.22 | 0.819 | 11.78 | 177.77 | 550.8 | 20.1 | 6576 | 0.405 |
| Kerosene pump-around | 490 | 38.6 | 0.831 | 11.83 | 180.00 | 388.2 | 14.6 | 42000 | 0.405 |
| Stripped diesel | 585 | 35.5 | 0.852 | 11.90 | 250.42 | 725.0 | 29.5 | 9627 | 0.405 |
| Atmospheric gas oil | 633 | 29.4 | 0.879 | 11.907 | 312.70 | 99.9 | 3.8 | 1280 | 0.468 |
| Crude bottoms | 659 | 18.33 | 0.941 | 11.580 | 424.81 | 2000 | 76.1 | 27436 | 0.600 |
| Naphtha splitter o/head | 289 | 65.90 | | 12.328 | 109.52 | 1385.8 | | 14485 | |
| Medium naphtha cond. | 130 | 65.90 | 0.717 | 12.328 | 109.52 | 1385.7 | 42.8 | 14477 | 0.484 |
| Naphtha splitter reflux | 130 | 65.90 | 0.717 | 12.328 | 109.52 | 681.4 | 20.3 | 7123 | 0.484 |
| Medium naphtha prod. | 130 | 65.90 | 0.717 | 12.328 | 109.52 | 704.2 | 21.5 | 7362 | 0.484 |
| Naphtha splitter b/toms | 401 | 46.66 | 0.794 | 11.721 | 142.38 | 7904.1 | 286.5 | 93475 | 0.413 |
| Reboiler feed | 401 | 46.66 | 0.794 | 11.721 | 142.38 | 7472.9 | 270.8 | 86484 | 0.413 |
| Heavy naphtha product | 401 | 46.66 | 0.794 | 11.721 | 142.38 | 436.2 | 15.63 | 4991 | 0.413 |

## Fig. 3.6a: Flowsheet Form of an Energy Balance Flowsheet (Vilbrandt & Dryden, 1959)

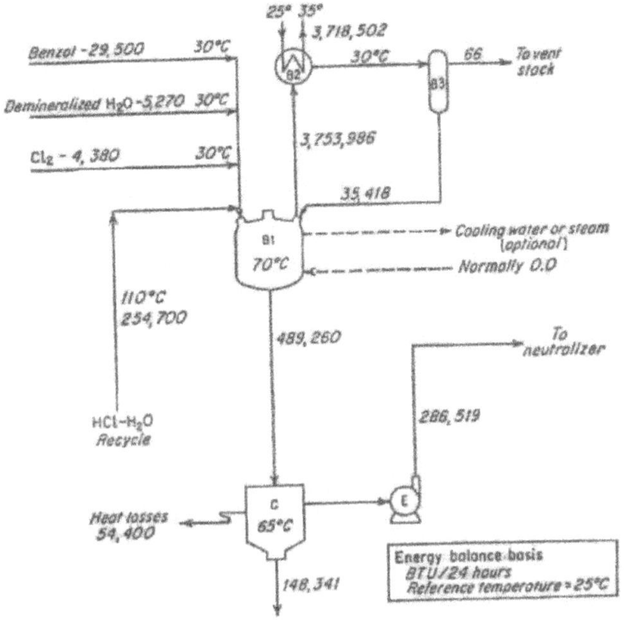

## Fig. 3.6b: Pictorial Form of an Energy Balance Sheet ( A Sankey Diagram (Austin, 1974))

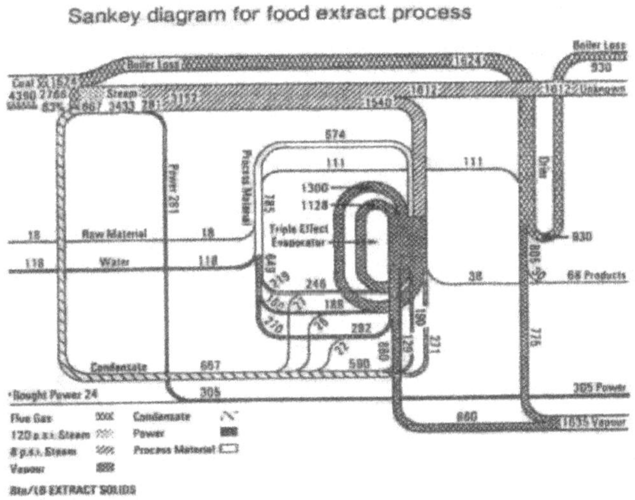

Sankey diagram for food extract process

## Fig. 3.7: Piping Flowsheet (Vilbrandt & Dryden, 1959)

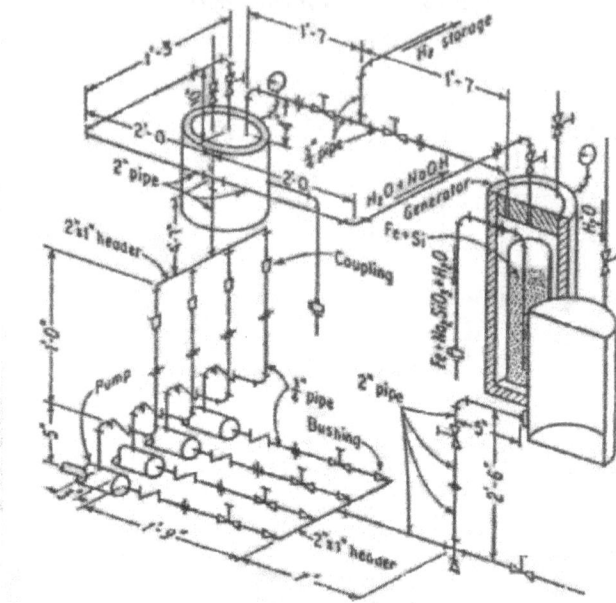

## Fig. 3.8:  An Instrumentation and Control Diagram

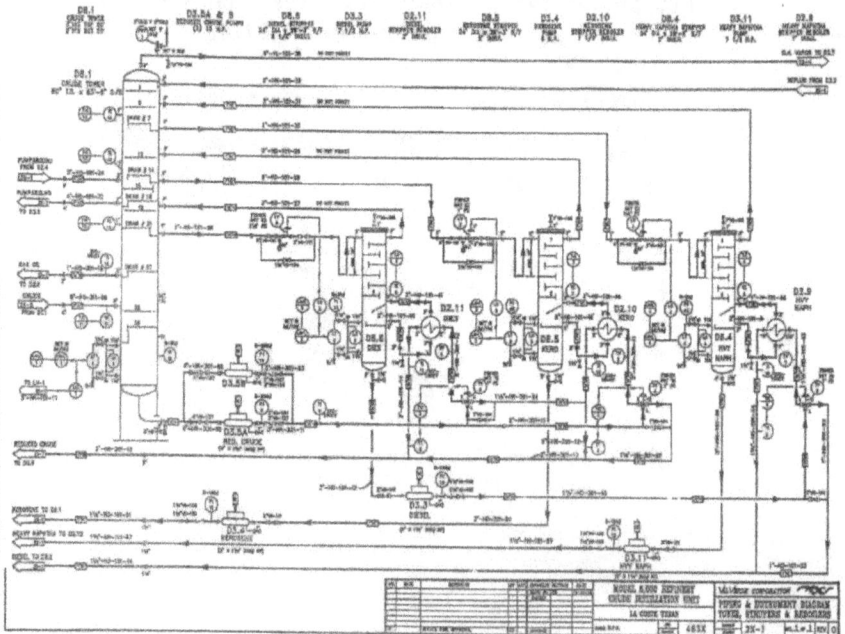

## 3.3: *Material and Energy Balances*

Material and energy balances are defined as the accounting of the material and energy flow and use in the process. This accounting enables us to determine:-

a) The raw material and energy requirements of a manufacturing process
b) The capacities of the various operating units of the process
c) The design of the various operating equipment or system.

To be able to do these, the laws of conservation of mass and energy, as treated in thermodynamics, are exceedingly, useful. Material and energy balance procedures will be treated in subsequent chapters. Thermodynamics and separation technology are not treated here, since they are treated, in detail, elsewhere.

## 3.4: *Equipment Design*

This is the translation of chemical engineering principles, associated with each operation, into pieces of hardware. It has four major components:

i) Chemical engineering design component. This determines the operating conditions of equipment as specified by process parameters. A typical example is the reactor where the reactor volume is determined for a given change in concentration or fractional conversion of feed, under given temperature and pressure conditions.

ii) Mechanical engineering design component. Here, the mechanical aspects of major equipment and units are determined and specified such as allowable pressures, temperatures, wall thicknesses, flange thicknesses, etc.

iii) Instrumentation and piping design component. The instrumentation and piping associated with each equipment are rationalised and specified in detail and provide reliable data for the overall plant instrumentation and piping design.

iv) Costing component. Every design must be evaluated to determine its cost and its cost contribution to the total plant or project cost.

Finally, every piece of equipment, whether designed from scratch or selected for purchase from a vendor, must have:

a).    Specifications, which give information, in a brief format, of all the
       important characteristics of the equipment,
b)     Drawings, which show, in sufficient detail, the pictorial/diagramatic
       representation of the equipment. These drawings are of many kinds
       though their usefulness depends on the state of equipment fabrication
       and/or installation technology.

### 3.4.1:  Types of Equipment Drawings

Drawings for equipment, designed and fabricated from scratch, are of
the following types:

i)    Design presentation sketches. These are representations of equipment
      using standardised chemical engineering equipment symbols, such as are
      listed in Austin, (1979).  They are used mostly in flowsheets of the
      complete plant or sections of it. The emphasis is on simplicity and ability
      to convey meaning just by looking at the sketch.

ii)   Design assembly/layout drawings. These are more detailed than
      presentation sketches and show essential dimensions and how the various
      parts fit each other. Codes, such as the British Standard (BS.974, 1553,
      1646, 5070) , the American National Standards Institute (ANSI. Y32.11,
      Y32.2.6, Z32.2.3, Z32.2.6) or the German standard codes, (DIN. 30600,
      40716) are used to designate standard parts, materials, finishes,
      tolerances and other information required by the draughtsman for
      producing detailed drawings. Note that codes represent either tested
      experience or some legal or safety requirement for the part designated by
      the code.

iii)  Detailed drawings. These show the engineering details of each part of the
      equipment drawn accurately to scale with principal dimensions and
      manufacturing instructions.

iv)   Shop assembly drawings. These show all the equipment and their parts,
      each properly coded so that they can be assembled and erected on site.

Shop assembly drawings involve further drawings which are of two
types:-

      (a) Outline assembly (general) drawings which show the exterior
          shape and principal dimensions of the equipment, overall clearance
          (headroom) requirements, anchoring bolt locations, piping and
          other instrument connections. These drawings make it possible to
          fit equipment, properly, into the plant layout

(b)  Equipment installation drawings. These show the details and the sequence to be followed in installing the equipment.

## 3.5: Equipment Specifications

After the material and energy balances, specifications of major items of equipment need to be determined, in sufficient detail, for accurate costing to be made. The designed equipment must conform to standard practice as specified in established codes such as the British Standard codes (B.S.), the American Society of Mechanical Engineers (A.S.M.E.), American Society for Testing Materials (A.S.T.M), etc.

The design is, usually, optimised with respect to purchase and operating costs. Mechanical design is included to deal with problems of temperature, pressure, corrosion, erosion, metal fatigue, etc. Equipment drawings are mandatory. Specifications for equipment can be preliminary, detailed or tailored to bidding purposes.

a)      Preliminary specifications are used for initial plant layout design or for pre-construction cost estimation. Typical items are listed in Table 3.2 and are, usually, developed by the user of the equipment.

b)  Detailed specifications. When a preliminary specification is sent to the vendor, he/she will return a detailed specification based on his/her own expertise in the field and on the information given in the preliminary specification. The accuracy of such a specification is related to the accuracy of the information supplied to the vendor. Table 3.3 lists items, normally, covered in a detailed specification. Sometimes, for proprietary reasons, the detailed estimates are better prepared by the buyer.

c). Specifications for competitive bidding. These are specifications given to suppliers of equipment to enable them submit a quotation or a bid. The specifications consist of only process data and such as to give enough information for the bid but not enough to suggest a particular equipment type. This way, individual suppliers are able to expose the buyer to the various levels and variety of design and technology.

**Table 3.2: Items, usually, listed in a Preliminary Equipment Specifications
Sheet (Vilbrandt & Dryden, 1959)**

| S/No. | Item |
|-------|------|
| 1 | Name and type of equipment, number required |
| 2 | Process materials, operating conditions |
| 3 | Design throughput and volume capacities |
| 4 | Dimensions, including design dimension such as surface area (for heat exchangers), volume (for reactors) or height (for packed towers), etc. |
| 5 | Recommended materials of construction |
| 6 | Piping requirements and other fittings |
| 7 | Instrumentation requirements |
| 8 | Utility requirements (electricity, electrical equipment and machines, steam, gas, compressed air, cooling water, etc.) |
| 9 | Construction details |
| 10 | Possible suppliers |
| 11 | Estimated operating labour |
| 12 | Estimated cost |
| 13 | Other notes and remarks |

**Table 3.3: Items, usually, listed in a Detailed Equipment Specifications
Sheet (Vilbrandt & Dryden, 1959)**

| S/No. | Item |
|-------|------|
| 1 | Name of equipment |
| 2 | Number required |
| 3 | Type and size |
| 4 | Materials of construction |
| 5 | Materials handled |
| 6 | Installation by whom |
| 7 | Delivery |
| 8 | Allowable variations |

**Fig.3.9: Examples of Standardised Chemical Plant Symbols (Austin, 1979) (Vilbrandt & Dryden, 1959)**

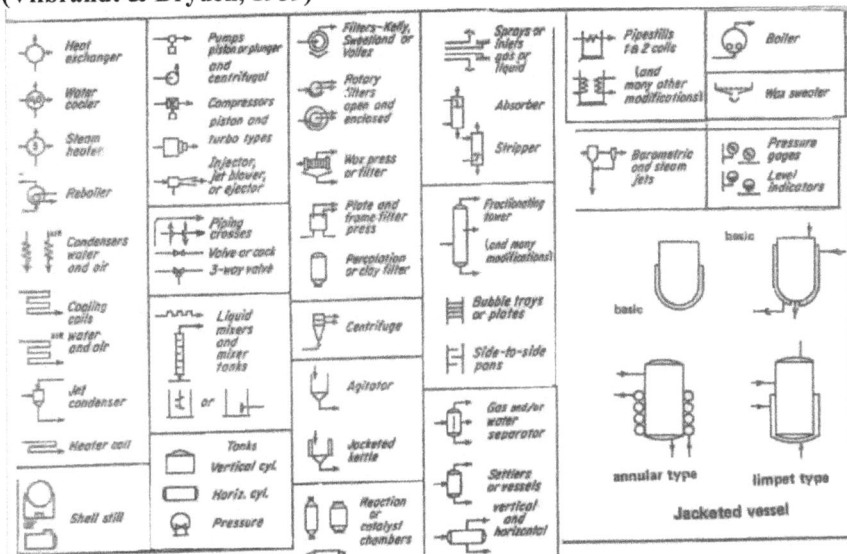

## References For Chapter Three

1  Austin D. G., *Chemical Engineering Drawing Symbols*; George Goodwin Ltd., London UK & Halsted Press, N.Y. U.S.A., 1979

2  Austin L. G., *Fuel Efficiency via the Mass Energy Balance*; ChemTech; Oct. 1974; pp 631-638; Am. Chem. Soc., Wash. D.C., U.S.A.

3  Filstrup P; *Handbook for the Meat Products Industry*; Alfa-Laval Slaughterhouse Byproducts Dept.; Titan Separator A/S Denmark, 1976

4  Rudd D. F., Powers G.J., and Siirola G.J.; *Process Synthesis*; Prentice Hall International; N.J., U.S.A.,1978

5  Scientific Design Co.; *The Scientific Design Story*; Scientific Design Co., Inc.; N. Y., U.S.A., 1968

6  Val Verde *6000 BPSD Crude Oil Fractionating Unit*; Chemex Inc.; California, U.S.A., 1999

7  Vilbrandt F. C., Dryden C. E.; *Chemical Engineering Plant Design*; 4th. Edition., International Student Edition; McGraw-Hill Book Co.; N.Y., U.S.A., 1959

105

# CHAPTER FOUR
# THE ELEMENTARY MATHEMATICS
# AND THERMODYNAMICS OF MATERIAL
# AND ENERGY BALANCES

Manufacturing in the chemical industry is all about consumption of materials and energy in order to obtain desirable products. Accounting for these enables us to measure of how effectively energy and materials have been used. The accounting for the inputs and outputs of material and energy, in a chemical process or plant, is what is, generally, referred to as material and energy balances. Since its development as a formal subject area in chemical process analysis and design around 1926 (Hougen, 1977), material and energy balancing has taken various forms and reached various levels of complexity.

The basic foundations of all material and energy accounting, however, are the laws of conservation of mass and energy. The systems to which these laws are applicable are, mainly, closed and open systems (in the thermodynamic sense). Such systems may be operating at steady state or unsteady state, though for manufacturing, steady state operations are preferred since they result in consistent production rate and quality. Unsteady state operation is, usually, studied because it gives an insight into what happens when a process is started up, closed down or has an unexpected disturbance or interruption.

Regardless of whatever operating state is to be analysed, standard definitions of process units and operations have been developed or have evolved over the years which can be used in most analyses of material and energy balances. These will be summarised later below. First, some basic definitions.

### 4.1: Mass Fraction

From the law of conservation of mass, it is clear that a system of N components, each component of mass, $m_i$, will have a total mass, $m_T$, given by

$$m_T = \sum_{i}^{N} m_i \qquad (4.1)$$

The mass fraction, $x_j$, of any component in the system is defined such that

$$x_i = \frac{m_i}{m_T} = \frac{m_i}{\sum\limits_{i}^{N} m_i} \qquad (4.2)$$

It is clear that

$$\sum\limits_{i}^{N} x_i = 1.0 \qquad (4.3)$$

## Illustrative Example 4.1

Derive expressions for the mass fraction of two components A and B, each of mass $M_A$ and $M_B$, respectively, in a system of total mass, $M_T$. Derive also, the relationships between them.

**Answer**

Total mass, $$M_T = \sum\limits_{i=1}^{2} M_i = M_A + M_B \qquad (1)$$

Mass fraction of A, $$x_A = \frac{M_A}{\sum\limits_{i=1}^{2} M_i} = \frac{M_A}{M_A + M_B} \qquad (2)$$

Mass fraction of B, $$x_B = \frac{M_B}{\sum\limits_{i=1}^{2} M_i} = \frac{M_B}{M_A + M_B} \qquad (3)$$

Then $$x_A + x_B = \frac{M_A}{M_A + M_B} + \frac{M_B}{M_A + M_B} = 1 \qquad (4)$$

Also $$x_A = 1 - x_B \quad and \quad x_B = 1 - x_A \qquad (5)$$

## 4.2: Mole Fraction

A system of N components, each of $n$ moles, will have a total number of moles, $n_T$, given by

$$n_T = \sum\limits_{i=1}^{N} n_i \qquad (4.4)$$

The mole fraction, $x_i$, of any component in the system is defined such that

$$x_i = \frac{n_i}{n_T} = \frac{n_i}{\sum\limits_{i=1}^{N} n_i} \qquad (4.5)$$

It is, also, true that

$$\sum_{i=1}^{N} x_i = 1.0 \qquad (4.3)$$

Since $n_i = m_i / FW_i$ where $FW_i$ = formula weight of component $i$, then

$$x_i = \frac{n_i}{n_T} = \frac{\dfrac{m_i}{FW_i}}{\displaystyle\sum_{i=1}^{N} \dfrac{m_i}{FW_i}} \qquad (4.4)$$

## Illustrative Example 4.2

An ethanol – water mixture was made up of 40 % by volume of ethanol at 15.5 C. What is the mole fraction of ethanol in the mixture?

## Answer

Consider 100 mls of this mixture at 15.5 C. Then, volume of ethanol = 40 ml; volume of water = 60 ml.  From data Tables we can find that the density of ethanol is 0.789 g/cc. Density of water is 1.00 g/cc. Molecular weight of ethanol is 46.07. That of water is 18.

Thus, moles of ethanol in the mixture;

$$n_{ethanol} = \frac{40 \; mls \; x \; density}{mol \; wt} = \frac{40 \, x \, 0.789}{46.07} = 0.685 \; gmols$$

Moles of water in the mixture;

$$n_{water} = \frac{60 \; mls \; x \; density}{mol \; wt} = \frac{60 \, x \, 1.00}{18} = 3.333 \; gmols$$

Hence mole fraction of ethanol

$$x_i = \frac{n_i}{n_T} = \frac{n_i}{\displaystyle\sum_{i=1}^{N} n_i} = \frac{0.685}{0.685 + 3.333} = 0.170 \quad Ans$$

Note that, from (4.3), the mole fraction of water, in the mixture, is $1 - 0.170 = 0.83$ and that the mole fractions of ethanol and water are different from their volume fractions.

## 4.3: Mass and Mole Ratios

These are the ratios of the mass or mole of, say, component, $i$, to, say, component, $j$, in the same system. Thus

$$Mass\ ratio,\ X_{ij} = \frac{m_i}{m_j} \qquad (4.6)$$

$$Mole\ ratio,\ \overline{X}_{ij} = \frac{n_i}{n_j} \qquad (4.7)$$

Generally, the mass or mole fraction, $x$, is related to the mass or mole ratio, $X$, as

$$x = \frac{X}{1+X} \qquad (4.8)$$

$$X = \frac{x}{1-x} \qquad (4.9)$$

### 4.4: Molarity and Molality

These terms are of particular interest to chemists but they are, also, very important in process design because the properties of solutions may only be available in chemical tables in those terms.

The *molarity* of a solute is the number of grams of it dissolved in 1000 ml of solution. The *molality* of a solute, on the other hand, is the number of grams of solute dissolved in 1000 grams of solvent.

Molarity and molality are not defined for solutes which are not soluble in a solvent or do not form a solution with it.

### 4.5: Pure Component Volume

The pure component volume, of a component gas in a mixture of gases, is the volume that component would occupy if it, alone, occupied the volume occupied by the mixture of gases, at the same temperature and pressure. This is different from the partial volume of a component gas which is the differential increase in volume when the component gas is added to the mixture.

### 4.6: The Law of Amagat or Leduc's Law

This law states that the total volume, $V_T$, occupied by a gaseous mixture, is equal to the sum of the pure component volumes, $V_i$, that is:

$$V_T = V_1 + V_2 + V_3 + ... V_N = \sum_{i=1}^{N} V_i \qquad (4.10)$$

For an ideal gas, the pure component volume $V_i = y_i$. $V_T$, where $y_i$ is the mole fraction of component $i$.

## Illustrative Example 4.3

100 kg of gas at 2 atm, 538 C, contain 57 kg $N_2$, 15 kg $O_2$, 20 kg $CO_2$ and 8 kg CO. Assuming ideal gas behaviour, determine the pure component volumes of $N_2$, $O_2$, $CO_2$, and CO.

## Answer

From the ideal gas law

$$PV = nRT \quad and \quad V = n\frac{RT}{P} \quad (1)$$

To get the value of R, the universal gas constant, consider the conditions of 1 kmol of an ideal gas at STP where $P_0 = 1$ atm, $V_0 = 22.414$ $m^3$/kmol and $T_0 = 273$ K. Then

$$R = \frac{P_0 V_0}{T_0} = \frac{1\,atm \times 22.414\,m^3\,/\,kmol}{273\,K} = 0.082\frac{atm.m^3}{kmol.K} \quad (2)$$

$$V = n\frac{RT}{P} = n\frac{0.082 \times (538 + 273)}{2}\,kmol.\frac{atm.m^3}{kmol.K}.\frac{K}{atm} = 33.251\,n, m^3 \quad (3)$$

Thus, from equation (3), the following Table gives the pure component volumes of the gases

| Component | Weight in Mixture, kg | Molecular Weight, kg | kmoles in Mixture | $V = n\frac{RT}{P}$ |
|-----------|------------------------|----------------------|-------------------|----------------------|
| $N_2$ | 57 | 28 | 2.04 | 67.83 |
| $O_2$ | 15 | 32 | 0.47 | 15.63 |
| $CO_2$ | 20 | 44 | 0.46 | 15.30 |
| CO | 8 | 28 | 0.29 | 9.64 |
| Total | 100 | | 3.26 | 108.40 |

Note that the total volume is equal to the sum of the pure component volumes, barring some rounding errors.

For liquids, the total volume is not equal to the sum of the pure component volumes because a shrinkage or volume increase can occur whenever pure component liquids are mixed.

Because volume expansions, or contractions, are different for different pure liquid components at different temperatures, volumetric compositions of

111

liquid mixtures can change with change in temperature.

## 4.7: Ideal Gas Volume

This concept is very useful for many purposes. It is found that 1 kmol of an ideal gas at standard temperature and pressure (STP) occupies 22.414 $m^3$. For example, since S.T.P. is equivalent to a temperature of 273 K, a pressure of 1 atm. or 1.013 x $10^5$ $N/m^2$ or 14.5 bar, and PV = nRT for an ideal gas, the ideal gas constant, R, and the mass of gas, at any condition, can be easily calculated, provided that the gas can be represented by the ideal gas equation.

### Illustrative Example 4.4

Calculate the ideal gas constant and the mass of 10 $m^3$ of carbon dioxide stored at 30 C and 1 atm pressure.

### Answer

The ideal gas constant is given, for one kmol of ideal gas, by

$$R = \frac{P_o V_o}{T_o} = \frac{1.013 \times 10^5 \times 22.414}{273} , \frac{N}{m^2} . \frac{m^3}{kmol} . \frac{1}{K}$$

$$= 8317 \frac{Nm}{kmol.K} = 8.317 \frac{kJ}{kmol.K} \quad Ans.$$

To calculate the mass of $CO_2$, recall from the ideal gas equation that the volume of $CO_2$ at STP is, since 30 C is equivalent to 303 K

$$V_o = \frac{T_o V_1}{T_1} = \frac{273 \times 10}{303} = 9.01 \, m^3/kmol$$

Since 1 kmol of $CO_2$ occupies 22.414 $m^3$ at STP, 9.01 $m^3$ of $CO_2$ at STP will have a mass of

$$\frac{9.01}{22.414} , \frac{m^3}{m^3} \frac{kmol}{} = 0.402, kmol = 0.402 \times 44, \frac{kmol}{} \frac{kg}{kmol} = 17.69 \, kg \quad Ans.$$

A more straightforward route would be to use PV = nRT from which

$$n = \frac{P_1 V_1}{RT_1} = \frac{1.013 \times 10^5 \times 10}{8317 \times 303} , \frac{N}{m^2} . \frac{m^3}{Nm} \frac{kmol.K}{} . \frac{1}{K} = 0.402 \, kmol \quad Ans.$$

## 4.8: Partial Pressure

In a mixture of gases, *the partial pressure*, of any of the component gases, is the pressure which that component would exert if it, alone, occupied the volume occupied by the mixture, at the same temperature and pressure.

## 4.9: Dalton's Law of Partial Pressures

This law states that the total pressure, exerted by a gas mixture, is equal
to the sum of the partial pressures of its component gases. For an ideal gas,
the total pressure, P, is given by $P = nRT/V$. For a mixture of ideal gases,
A, B and C, the partial pressures of the individual gases are, also, given
as:

$$p_A = n_A \frac{RT}{V} \tag{4.11}$$

where $n_A$ = moles of A

$$p_B = n_B \frac{RT}{V} \tag{4.12}$$

where $n_B$ = moles of B

$$p_C = n_C \frac{RT}{V} \tag{4.13}$$

where $n_C$ = moles of C.

By Dalton's law of partial pressures, the total pressure, $P_T$, of this mixture
of gases becomes

$$P_T = p_A + p_B + p_C = (n_A + n_B + n_C)\frac{RT}{V} \tag{4.14}$$

If we designate the mole fraction of the component gas, A, as $y_A$ then from
(4.5)

$$y_A = \frac{n_A}{n_A + n_B + n_C} \tag{4.15}$$

It can be seen, from (4.11), (4.14) and (4.15), that

$$y_A = \frac{n_A}{n_A + n_B + n_C} = \frac{p_A}{P_T} \quad or \quad p_A = y_A P_T \tag{4.16}$$

That is, the partial pressure of component A, in a gas mixture, is equal to its
mole fraction multiplied by the total pressure of the mixture. As for
component A, so also for each of the component gases in the mixture.

### Illustrative Example 4.5

100 kg of gas at 2 atm, 538 C, contain 57 kg $N_2$, 15 kg $O_2$, 20 kg $CO_2$
and 8 kg CO. Assuming ideal gas behaviour, determine a) the mole
fractions, b) the partial pressures of the component gases, c) the mean
molecular weight of the mixture and d) the volume occupied by the

mixture. Take the formula weights to be $N_2$, 28; $O_2$, 32; $CO_2$, 44 and CO, 28. 1 kmol of an ideal gas, at STP, occupies 22.414 $m^3$.

**Answer**

The computations for parts a) and b) may be done more compactly as in the Table below.

| Component | Weight in Mixture, kg | Molecular Weight, kg | kmols in Mixture | Mole Fraction | Partial Pressure, atm |
|---|---|---|---|---|---|
| $N_2$ | 57 | 28 | $\dfrac{57}{28}$ $= 2.036$ | $\dfrac{2.036}{3.246}$ $= 0.627$ | $0.627 \times 2$ $= 1.254$ |
| $O_2$ | 15 | 32 | $\dfrac{15}{32}$ $= 0.469$ | $\dfrac{0.469}{3.246}$ $= 0.145$ | $0.145 \times 2$ $= 0.290$ |
| $CO_2$ | 20 | 44 | $\dfrac{20}{44}$ $= 0.455$ | $\dfrac{0.455}{3.246}$ $= 0.140$ | $0.140 \times 2$ $= 0.280$ |
| CO | 8 | 28 | $\dfrac{8}{28}$ $= 0.286$ | $\dfrac{0.286}{3.246}$ $= 0.088$ | $0.088 \times 2$ $= 0.176$ |
| **Total** | **100** | | **3.246** | **1.000** | **2.000** |

c). To determine the mean molecular weight, we know that each components contributes according to its weight and mole fraction. Thus a Table can be constructed as shown below

| Component | Molecular Weight, kg | Mole Fraction | Molecular Weight Contribution, kg |
|---|---|---|---|
| $N_2$ | 28 | 0.627 | $0.627 \times 28 = 17.556$ |
| $O_2$ | 32 | 0.145 | $0.145 \times 32 = 4.640$ |
| $CO_2$ | 44 | 0.140 | $0.140 \times 44 = 6.160$ |
| CO | 28 | 0.088 | $0.088 \times 28 = 2.464$ |
| **Total** | | | **30.820** |

Note also that the mean molecular weight can be estimated from dividing the total weight of the mixture by its total equivalent moles. That is

$$Mean \quad molecular \ weight = \frac{100}{3.246} = 30.807$$

d) To calculate the volume occupied by the gas mixture, note that 538 C is equivalent to $538 + 273 = 811$ K and that 100 kg is equivalent to $100/30.82 = 3.245$ kmol of gas.

Since $\quad \dfrac{P_1 V_1}{T_1} = \dfrac{P_0 V_0}{T_0}$ and $P_1 = P_0 = 1$ atm,

then $\qquad V_1 = \dfrac{T_1 V_0}{T_0} = \dfrac{811 \times 22.414}{273} = 66.59 \ m^3 / kmol$

Total volume occupied $= 66.59 \times 3.246 = 216.15 \ m^3$. Ans

Alternatively, $PV = nRT$ or

$$V_1 = \frac{n \, R \, T_1}{P_1} = \frac{3.246 \times 8314 \times 811}{1.013 \times 10^5} , \frac{kmol}{} \cdot \frac{N \, m}{kmol.K} \cdot \frac{K}{} \cdot \frac{m^2}{N} = 216.06 \, m^3. \quad Ans.$$

## 4.10: Concentration in the Gas Phase

Concentration is, generally, defined as mass per unit volume of which it is part of as opposed to the ratio of mass to volume to which it is not a part. For an ideal gas, therefore, it follows that concentration, $C_A$ of a component, say A, will be given by

$$C_A = \frac{n_A}{V} = \frac{P_A}{RT} \qquad (4.17)$$

where $C_A$ has units of moles per unit volume. Often, when very large quantities of material are involved, the kilomole, abreviated kmol is used and often with the volume in $m^3$. The kmol is the mass of the compound in kilograms divided by the formula weight. For example, 100 kg. of KOH is $100/56 = 1.79$ kmol.

In addition to the brevity achieved with the use of kilomoles, the use of moles in process calculations has the advantage of being in conformity with the ratios (mole ratios) in which chemical reactions take place.

## 4.11: Density

Density is a different type of measure of mass per unit volume of which it is part. While concentration aims to describe the extent of distribution

115

of mass within a given volume, density expresses this as the ratio of the weight of the given volume to its volume.

For most pure liquids, except those that contain halogens or heavy atoms, the mass density lies between 800 and 1000 kg/m³. The mass density must be distinguished from the molar density of which the latter value may be estimated, at the normal boiling point, from

$$\rho = \frac{M}{V}$$ (4.18)

where M is the molecular weight and V is the molar volume

### 4.11.1: Density of Liquid Mixtures

For liquid mixtures the mass density of the mixture may be estimated, when measured values are not available, from

$$\rho_{mixture} = \sum x_i \, \rho_i$$ (4.19)

where $x_i$ and $\rho_i$ are the weight fraction and mass density, respectively, of component *i*.

**Illustrative Example 4.6**

Calculate the mass density of a mixture of methanol and water at 20 C if methanol is 40 % by weight of the solution. Take the mass density of water and methanol at 20 C to be 998.2 kg/m³ and 791.2 kg/m³, respectively.

**Answer**

$$Density\ of\ mixture = \sum_{2} x_i . \rho_i = 0.6 \, x \, 998.2 + 0.4 \, x \, 791.2 = 915.4 \, kg/m^3 \quad Ans.$$

The problem may also be solved another way if the mixture is assumed to be ideal. Thus, on the basis of 1000 kg of mixture

$$Volume\ of\ water = 0.6 \, x \, \frac{1000 \ kg}{998.2} . \frac{m^3}{kg} = 0.601 \, m^3$$

$$Volume\ of\ methanol = 0.4 \, x \, \frac{1000 \ kg}{791.2} . \frac{m^3}{kg} = 0.506 \, m^3$$

$$Total\ volume\ of\ water = 0.601 + 0.506 = 1.107 \, m^3$$

$$Density \ of \ mixture = \frac{1000}{1.107} \frac{kg}{m^3} = 903.3 \ kg/m^3 \quad Ans$$

The experimental value is 934.5 kg/m³. Thus the error in either method of estimation is, for equation (4.2)

$$Percent \ error = 100 \ x \frac{(934.5 - 915.4)}{934.5} = 2.04 \%$$

and for the assumption of an ideal solution

$$Percent \ error = 100 \ x \frac{(934.5 - 903.3)}{934.5} = 3.34 \%$$

Although there is not much difference between the two, it would appear that the assumption of an ideal solution gives the less accurate result for this mixture.

<u>Variation of Density with Temperature</u>

The variation of density with temperature, at constant total pressure, is, usually, given as

$$\rho_T = \rho_0 \left(1 + \beta \Delta T\right) \tag{4.20}$$

where $\rho_T$, $\rho_0$, are the densities at the temperature T and at some reference temperature, respectively, and $\Delta T$ is the temperature difference between T and the reference temperature.

For non-polar liquids

$$\beta = \frac{0.04314}{\left(T_C - T\right)^{0.641}} \tag{4.21}$$

where $\beta$ = coefficient of thermal expansion, $T_C$ = critical temperature, K, T = temperature, K

### 4.11.2: Density of Pure Gases

For an ideal gas or a real gas under a total pressure under 20 bar, the ideal gas equation is generally applicable to some degree of accuracy. For such a gas,

$$Specific \ volume = \frac{1}{Density} = \frac{RT}{P} \tag{4.22}$$

where P and T are absolute values.

For greater accuracy for real gases, an adjusted ideal gas equation is used. Such an equation is

$$PV = Zn \ RT \tag{4.23}$$

117

where $Z$ = compressibility factor. The compressibility factor is estimated from the reduced temperature, $T_R$, and pressure, $P_R$ of the gas. The form of variation of $Z$ with $P_R$ and $T_R$ is illustrated in the compressibility chart of Figure 4.1 below.

To determine compressibility factors for gas mixtures, the reduced properties used are those summed from the contribution of each component. Thus

$$P_C = \sum_i y_i P_{C_i} \qquad (4.24)$$

$$T_C = \sum_i y_i T_{C_i} \qquad (4.25)$$

**Fig. 4.1: Compressibility Factors for Pure Gases (Sinnott, 1983)**

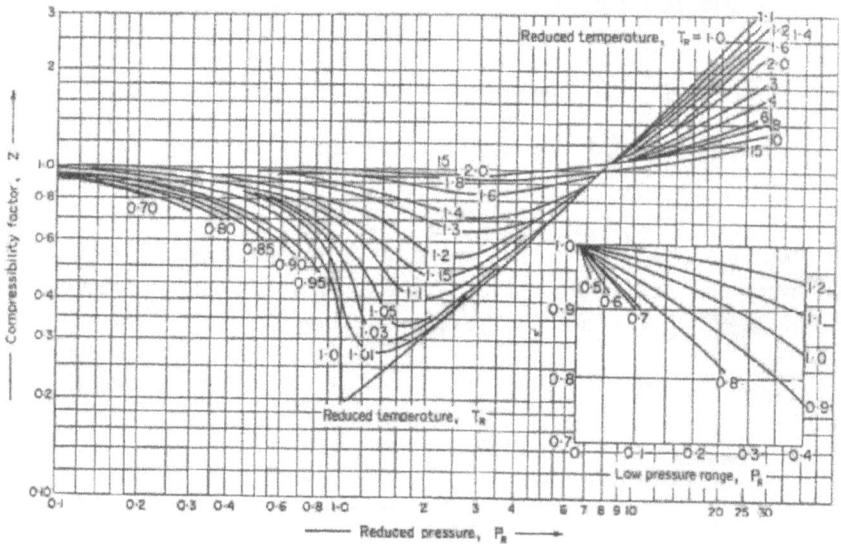

## 4.12: Specific Heat Capacity

The best values of heat capacity to use are those obtained experimentally. When these are not available, the following methods of estimating them may be used.

Estimating the Specific Heat Capacity of Solids and Liquids

Kopp's law may be used. This law states that the heat capacity of the solid or liquid compound may be obtained from the heat capacities of the elements of which it is composed. Thus

118

$$Cp = \frac{\sum_i \left(moles\ of\ element\ x\ heat\ capacity\ of\ element\ per\ mol\right)}{\sum_i molecular\ weight\ of\ element} \qquad (4.26)$$

A typical set of values of the heat capacity of some elements, at room temperature, is shown in Table 4.1.

**Table 4.1: Specific Heat Capacities of some Common Elements**

| Element | Solids, kJ/kmol.K | Liquids, kJ/kmol.K | Element | Solids, kJ/kmol.K | Liquids, kJ/kmol.K |
|---------|-------------------|--------------------|---------|-------------------|--------------------|
| C | 7.5 | 11.7 | Si | 15.9 | 24.3 |
| H | 9.6 | 18.0 | O | 16.7 | 25.1 |
| B | 11.3 | 19.7 | F | 20.9 | 29.3 |
| P & S | 22.6 | 31.0 | All Others | 26.0 | 33.5 |

## Illustrative Example 4.7

It was not possible to obtain the specific heat capacity of urea from standard tables. Estimate its value.

**Answer**

The molecular formula of urea, a colourless, prismatic solid, is $CH_4N_2O$, molecular weight, 60. We can construct a Table of the heat capacities of its elements as shown below

| Element | Atomic Weight | Moles | Heat Capacity, kJ/kmol.K | Moles x Heat Capacity, kJ/K |
|---------|---------------|-------|--------------------------|------------------------------|
| C | 12 | 1 | 7.5 | 7.5 |
| H | 1 | 4 | 9.6 | 38.4 |
| N | 14 | 2 | 26 | 52 |
| O | 16 | 1 | 16.7 | 16.7 |
| Total | | | | 114.6 |

Thus, from equation 4.26

$$Specific\ heat\ capacity\ of\ urea = \frac{114.6}{60} = 1.91\,kJ/kg.K \quad Ans$$

Note that the experimental value is 1.55 kJ/kg.K. This makes the estimated value +23.2% in error.

## Estimating the Specific Heat Capacity of Organic Liquids

For this, the group contribution method of Chueh and Swanson is used. This method is based on the same principles as the Kopp's method except that organic chemical groups, rather than elements, are used. Typical values are listed in Table 4.2.

**Table 4.2: Specific Heat Capacities of some Organic Chemical Groups**

| Alkanes | Value, kJ/kmol.K | Halogens | Value, kJ/kmol.K |
|---|---|---|---|
| - CH$_3$ | 36.84 | - Cl ($1^{ST}$ or $2^{ND}$ on a Carbon) | 36.01 |
| - CH$_2$ - | 30.40 | - Cl ($3^{RD}$ or $4^{TH}$ on a Carbon) | 25.12 |
| - CH - | 20.93 | - Br | 37.68 |
| $\begin{array}{c} \| \\ - C - \\ \| \end{array}$ | 7.37 | | |
| $\begin{array}{c} O \\ \| \| \\ - C - O - \end{array}$ | 60.71 | **Nitrogen** | **Value, kJ/kmol.K** |
| - CH$_2$OH | 73.27 | $\begin{array}{c} H - N - \\ \| \\ H \end{array}$ | 58.62 |
| $\begin{array}{c} \| \\ - CHOH \\ \| \end{array}$ | 76.20 | $\begin{array}{c} H \\ \| \\ - N - \\ \| \end{array}$ | 43.96 |
| - COH | 111.37 | $\begin{array}{c} \| \\ - N - \end{array}$ | 31.40 |
| - OH | 44.80 | - N = (In a ring) | 18.84 |
| - ONO$_2$ | 119.32 | - C $\equiv$ N | 58.70 |
| **Olefins** | | **Sulphur** | **Value, kJ/kmol.K** |
| =CH$_2$ | 21.77 | - SH | 44.80 |
| $\begin{array}{c} \| \\ =CH \\ \| \end{array}$ | 21.35 | - S - | 33.49 |
| $\begin{array}{c} \| \\ =C - \end{array}$ | 15.91 | - F | 16.75 |
| | | - I | 36.01 |
| | | **Hydrogen** | **Value, kJ/kmol.K** |
| | | H- ( formic acid, formals, HCN, etc) | 14.65 |

| In a Ring | | Alkyne | Value, kJ/kmol.K |
|---|---|---|---|
| $-CH=$ | 18.42 | $-C\equiv H$ | 24.70 |
| $-C=$ or C $-$ | 12.14 | $-C\equiv$ | 24.70 |
| $- CH=$ | 22.19 | | |
| $- CH_2 -$ | 25.96 | | |

## Illustrative Example 4.8

Estimate the specific heat capacity of ethyl bromide at 20 C

## Answer

The structural formula of ethyl bromide is

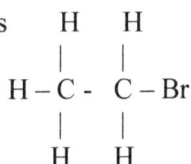

$$H - C - C - Br$$

with H H on top and H H on bottom.

We can construct a Table of the heat capacities of its constituent chemical organic groups as shown below

| Group | Contribution, kJ/kmol.K | Number | Sub-total, kJ/K |
|---|---|---|---|
| $- CH_3$ | 36.84 | 1 | 36.84 |
| $- CH_2 -$ | 30.4 | 1 | 30.4 |
| Br | 37.68 | 1 | 37.68 |
| | | Total | 104.92 |

Since the molecular weight is 109,

$$Specific\ heat\ capacity\ of\ ethyl\ bromide = \frac{104.92}{109} = 0.96\ kJ/kg.K \quad Ans$$

Note that the experimental value is 0.90 kJ/kg.K. This makes the estimated value +6.7% in error.

## Addition Rule

The addition rule makes it possible to use the method of Chueh and

121

Swansin in more complicated organic group situations. The rule, simply, adds 18.84 for any carbon group

- which is joined by a single bond to a carbon group connected by a double or triple bond with a third carbon group
- as many times as this occurs

Exceptions

a). No addition for – CH₃

b). For a – CH₂ – group meeting criteria, add 10.47 not 18.84

c). For a – CH₂ – group meeting criteria more than once, add 10.47 first time and 18.84 in subsequent ones

d). No additions for any carbon group in a ring

**Illustrative Example 4.9**

Estimate the specific heat capacity of chlorobutane at 20 C using Chueh's and Swanson's method.

**Answer**

$$CH_2 = C - CH = CH_2$$

The structural formula of chlorobutane is

$$|$$

$$Cl$$

A Table of its group contributions is shown below. Its molecular weight is 88.5.

| Group | Contribution, kJ/kmol.K | Number | Addition Rule, kJ/kmol.K | Total, kJ/K |
|-------|-------------------------|--------|--------------------------|-------------|
| = CH₂ | 21.77 | 2 | - | 43.54 |
| = C – | 15.91 | 1 | 18.84 | 34.75 |
| = C – H | 21.35 | 1 | 18.84 | 40.19 |
| - Cl | 36.01 | 1 | - | 36.01 |
| | | Totals | | 154.49 |

Thus

$$Specific\ heat\ capacity\ of\ chloro\ bromide = \frac{154.49}{88.5} = 1.75\,kJ/kg.K \quad Ans$$

The experimental value is 1.888 kJ/kg.K. This makes the estimated value -7.3% in error.

Note that

- liquid specific heats do not vary much with temperature at $T_R < 0.7$.
- for liquid mixtures, $Cp = \sum_i x_i \, Cp_i$  (4.27)

- for dilute solutions in water, use that of water.

## Estimating the Specific Heat Capacity of Gases

The specific heat capacity of most gases can be estimated from the formula

$$Cp = A + BT + CT^2 + DT^3 \qquad (4.28)$$

The values of A, B, C and D are usually available for the more common gases (see the Appendix)

For a mixtures of gases, the heat capacity of the mixture is given by equation (4.27)

## Use of Mean Heat Capacity

Sometimes it is more convenient, over a range of temperatures, say $T_1$ to $T_2$, to use a mean heat capacity. In such situations, the mean heat capacity, $Cp_m$ is

$$Cp_m = \frac{\int_{T_1}^{T_2} Cp\,dT}{\int_{T_1}^{T_2} dT} \qquad (4.29)$$

With the given expression for the heat capacity of the gas, this is equivalent to

$$Cp_m = \frac{a(T - T_{ref}) + \dfrac{b}{2}(T^2 - T_{ref}^2) + \dfrac{c}{3}(T^3 - T_{ref}^3) + \dfrac{d}{4}(T^4 - T_{ref}^4)}{T - T_{ref}} \qquad (4.30)$$

If we work in degrees Centigrade, we can simplify matters by choosing $T_{ref} = 0$. Then

$$Cp_m = a + \frac{b}{2}T + \frac{c}{3}T^2 + \frac{d}{4}T^3 \qquad (4.31)$$

123

## 4.13: Enthalpy

Enthalpy, H, is defined as the heat change in a thermodynamic process at constant total pressure. Thus

$$\Delta H = \int Cp\, dT \tag{4.32}$$

If there is no phase change during this process, the enthalpy may, also, be referred to as sensible heat. If the mean heat capacity is used, then

$$\Delta H = Cp_{m_2}(T_2 - T_{ref}) - Cp_{m_1}(T_1 - T_{ref}) \tag{4.33}$$

where $Cp_{m1}$ and $Cp_{m2}$ are the mean heat capacities evaluated between $T_1$ and $T_{ref}$ and $T_2$ and $T_{ref}$, respectively. If $T_{ref} = 0$, then

$$\Delta H = Cp_{m_2}T_2 - Cp_{m_1}T_1 \tag{4.34}$$

If the variation of Cp with T is not very great, equation (4.34) may be replaced by

$$\Delta H = Cp_m(T_2 - T_1) \tag{4.35}$$

where $Cp_m$ is the mean heat capacity evaluated between $T_1$ and $T_2$. These expressions turn out to be very useful because the heat capacity values that are available in Tables are, often, mean values.

### Illustrative Example 4.10

Calculate the enthalpy change when 1 mol of $CO_2$ at 200 C is heated to 800 C, given that the mean heat capacities of $CO_2$ at 200 C and 800 C are 40.15 and 47.94, J/mol. C, respectively.

### Answer

$$\Delta H = Cp_{M_2}t_2 - Cp_{M_1}t_1 = 47.94 \times 800 - 40.15 \times 200 = 30,322\, J/mol \quad Ans$$

When there is phase change, the total enthalpy change is the sum of the sensible heat and the heat of the phase change, known as the heat of transition, which may be evaporation, condensation, freezing, sublimation, etc. If the heat of transition is designated as $\lambda$, then, per unit mass of substance,

$$\Delta H = \int_{T_{ref}}^{T_{transition}} Cp\, dT + \lambda \tag{4.36}$$

### Illustrative Example 4.11

Calculate the enthalpy change when 1 kmol of water at 30 C is completely evaporated at 100 C, given that the mean heat capacities of water at 30 C and 100 C are 4179 and 4219 J/kg. C, respectively. Take

124

the latent heat of water at 100 C as 2,256,700 J/kg

**Answer**

Assume complete evaporation at 100 C.
$$\Delta H = Cp_{m_2} T_2 - Cp_{m_1} T_1 + \lambda = 4219 \times 100 - 4179 \times 30 + 2256700$$
$$= \frac{2,553,230}{18} = 141,846 \, J \, / \, kmol \quad Ans.$$

## *4.14: The Second Law of Thermodynamics in Process Design*

The second law of thermodynamics plays key roles in chemical reactions, especially in indicating how spontaneously they are likely to occur, the direction of reaction and what their equilibrium is likely to be.

It has many definitions which, on reflection, boil down to the same theme. When defined in terms of thermal energy, the second law of thermodynamics postulates that in any system, thermal energy is always degraded, during a process, into entropy. In terms of work energy, the second law states that not all energy put into a system to do work can be converted into work. Some of this energy must be rejected to the surroundings.

In terms of natural, versus man made, processes, the second law states that all natural processes are irreversible. In terms of the stability, spontaneity or equilibrium of processes, the second law states that all natural processes tend to move to a state of higher disorder (entropy) or to a state of the lowest energy.

Consequently and for process systems, the second law of thermodynamics enables us to define the conditions of thermodynamic equilibrium much more exactly and usefully. These conditions are

1. There must be thermal equilibrium. That is, there must be no temperature differences between elements in the system or between the system and its surroundings.
2. There must be mechanical equilibrium. That is, there are no unbalanced forces between elements in the system or between the system and its surroundings.
3. There must be chemical equilibrium. That is, there are no longer

changes in structure or composition in the system. Chemical equilibrium exists when the system's Helmholtz's free energy, A, at constant T and V or its Gibb's free energy, G, at constant T and P, is constant. That is $\Delta A = 0$ at constant T and V or $\Delta G = 0$ at constant T and P.

These conditions of thermodynamic equilibrium enable us to develop further conditions for phase and chemical equilibrium in process systems, especially those defined by their temperature, T, volume, V and pressure, P, otherwise known as PVT systems.

Usually, our interest in phase and chemical equilibrium rests on our need to determine, for non-reacting systems

a) The heat transferred between any two phases within a heterogeneous system
b) The mass transfer, of any component in the system, across a phase boundary (Prausnitz, 1969)
c) Whether there is, and the extent of, a displacement of the phase boundary

and additionally, for reacting systems,

d) the equilibrium composition of the reacting systems and the effects of temperature and pressure on them. This, usually, results in an equilibrium constant

### 4.15: Thermodynamic Descriptions of Phase Equilibrium

The systems of interest in process design are either homogenous or heterogenous and may be either closed or open systems.

For homogenous systems, Prausnitz (1969) has summarized standard mathematical descriptions of phase equilibrium as listed in Tables 4.3 and 4.4. The dependent variables of interest are the internal energy, U, the enthalpy, H, the entropy, S, and the chemical potential, $\mu$. The independent variables are, usually, the measurable properties of the system such as temperature, volume, pressure, composition, etc.

For heterogenous systems, phase equilibrium is described by the Gibbs-Duhem equation (which describes equilibrium also in open homogenous systems), the phase rule and fugacity, activity and activity coefficients.

126

**Table 4.3: Thermodynamic Relations for Closed Homogenous Systems (Prausnitz, 1969)**

| Fundamental Equations | Equilibrium Criteria | Mathematical Identities |
|---|---|---|
| $dU = TdS - PdV$ | $dU\|_{S,V} \leq 0$ | $\left(\dfrac{\partial U}{\partial S}\right)_V = \left(\dfrac{\partial H}{\partial S}\right)_P$ |
| $dH = TdS + VdP$ | $dH\|_{S,P} \leq 0$ | $\left(\dfrac{\partial U}{\partial V}\right)_S = -P = \left(\dfrac{\partial A}{\partial V}\right)_T$ |
| $dA = -SdT - PdV$ | $dA\|_{T,V} \leq 0$ | $\left(\dfrac{\partial H}{\partial P}\right)_S = V = \left(\dfrac{\partial G}{\partial P}\right)_T$ |
| $dG = -SdT + VdP$ | $dG\|_{T,P} \leq 0$ | $\left(\dfrac{\partial A}{\partial T}\right)_V = -S = \left(\dfrac{\partial G}{\partial T}\right)_P$ |

The Gibbs-Duhem equation is stated as

$$dG = -SdT + VdP + \sum_i \mu_i \, dn_i \qquad (4.37)$$

where $n_i$ is the number of moles of component $i$.

The phase rule sets the limits to the number of phases which can be in equilibrium with each other in the same system. It also sets limits on the phases which are compatible with each other. A statement of the phase rule is

$$F = C + 2 - P \qquad (4.38)$$

F is the number of degrees of freedom, C is the number of species or components present in the system and P is the number of phases formed.

The chemical potential, $\mu$, for an ideal gas is derived to be

$$\mu_i = \mu_i^o + RT \ln \frac{P_i}{P_i^o} \qquad (4.39)$$

where $P_i$ is its partial or vapour pressure at the temperature T and $P_i^o$ is this pressure at its standard or reference state. $\mu_i^o$ is the chemical potential at its standard or reference state.

This derivation can be made more general to also describe solids, liquids or gases, pure or mixed, ideal or non-ideal, by the use of the

127

term, fugacity.

**Table 4.4: Thermodynamic Relations for Open Homogenous Systems (Prausnitz, 1969)**

| Fundamental Equations | Equilibrium Criteria | Mathematical Identities |
|---|---|---|
| $dU = TdS - PdV + \sum_i \mu_i \, dn_i$ | $dU\big|_{S,V,n} \leq 0$ | $\mu_i = \left(\dfrac{\partial U}{\partial n}\right)_{S,V}$ |
| $dH = TdS + VdP + \sum_i \mu_i \, dn_i$ | $dH\big|_{S,P,n} \leq 0$ | $\mu_i = \left(\dfrac{\partial H}{\partial n}\right)_{S,P}$ |
| $dA = -SdT - PdV + \sum_i \mu_i \, dn_i$ | $dA\big|_{T,V,n} \leq 0$ | $\mu_i = \left(\dfrac{\partial A}{\partial n}\right)_{T,V}$ |
| $dG = -SdT + VdP + \sum_i \mu_i \, dn_i$ | $dG\big|_{T,P,n} \leq 0$ | $\mu_i = \left(\dfrac{\partial G}{\partial n}\right)_{T,P}$ |

By defining fugacity, $f_i$, as

$$\frac{f_i}{y_i P_i} \to 1 \quad as \quad P_i \to 0 \tag{4.40}$$

where $y_i$ is the mole fraction, equation (4.20) becomes

$$\mu_i = \mu_i^o + RT \ln \frac{f_i}{f_i^o} \tag{4.41}$$

$f_i^o$ is the fugacity at the standard or reference state. $f_i / f_i^o$ is defined as the activity, $a_i$. That is,

$$a_i = \frac{f_i}{f_i^o} \tag{4.42}$$

The activity coefficient, $\gamma_i$, of component $i$, is, thus defined as

$$a_i = \gamma_i \, C_i \tag{4.43}$$

where $C_i$ is the molar concentration.

Equations (4.37) to (4.43) form the basis of most analysis and prediction of the properties of solutions and of solid-liquid, liquid-liquid and liquid - vapour equilibria, especially, in the prediction of K-values. Typical examples of analytical and prediction methods, in use, are the Raoult's and Henry's laws, the Antoine equation, Clausius - Clayperon equation and many others.

## 4.15.1: *Ideal Solutions and Mixtures*

An ideal solution is a theoretical abstraction, just like the concept of an ideal gas. This concept has proved to be very useful in many practical situations in which the properties and behaviour of liquids need to be either understood or exploited.

An ideal solution or mixture is that in which the enthalpy of solution or of mixing is zero. The activity coefficient (which measures deviation from ideality) of such a mixture is equal to one.

The difference between the ideality of solutions and that of gases is that intermolecular interactions in liquids are strong and cannot simply be neglected as they can be for ideal gases. For liquids, the mean strength of the interactions is assumed to be the same between all the molecules of the solution.

An ideal liquid mixture is defined, formally, as a mixture that satisfies

$$f_i = x_i f_i^0 \qquad (4.44)$$

where $f_i$ and $x_i$ are the fugacity and mole fraction in the liquid phase, respectively, of component $i$ and $f_i^0$ is the fugacity of $i$ as a pure substance.

Vapour Pressure, at any Temperature

This is the equilibrium or saturation pressure exerted by escaping molecules from the surface of a liquid. It is a function of only temperature except at very high pressures.

Normal Boiling Point of a Liquid

This is the temperature at which the saturation vapour pressure of the liquid is equal to atmospheric pressure (760 mmHg.). In a mixture of gases and vapours, the partial pressure of any component, at any temperature, is regarded as its equilibrium vapour pressure at that temperature.

Saturation Temperature or Dew Point

This is the temperature at which the vapour condenses to a liquid

Superheated Vapour

This is the vapour whose partial pressure is less than its equilibrium or saturation vapour pressure. Such a vapour is unsaturated.

Degrees of Superheat

This is the amount by which the actual temperature of a vapour exceeds its saturation temperature.

### 4.15.1.2: Prediction of Vapour Pressure of Pure Substances: the Antoine Equation

A very accurate and widely used equation for estimating the vapour pressure of pure substances is the Antoine's equation. This equation takes the form

$$\ln p = A - \frac{B}{T + C} \tag{4.45}$$

where $p$ is the vapour pressure in mm Hg, A, B, C are constants and T is the absolute temperature.

### Illustrative Example 4.12

The Antoine equation for the prediction of the vapour pressure of pure substances is given by equation (4.45). If $A = 18.3036$, $B = 3816.44$ and $C = -46.13$ for water, determine the vapour pressure of water at 100 C.

### Answer

We can substitute the given values of the constants in equation (4.45) to get

$$\ln p = A - \frac{B}{T + C} = 18.306 - \frac{3816.44}{373 - 46.13} = 6.6303$$

That is $p = 757.7$ mm Hg. The correct value is 760 mm Hg, an error of 0.3%.

### 4.15.1.3: Predicting Vapour Pressures of Components of Liquid Mixtures: Raoult's Law

Every pure liquid exerts a vapour pressure, the extent of which depends on the liquid temperature and the ambient pressure. Raoult's law, simply, states that, for a liquid mixture of volatile components, the equilibrium vapour pressure exerted by a component in the solution is proportional to the mole fraction of that component in the solution. That is:

$$P_A = x_A P_A^0 \qquad (4.46)$$

where $P_A$ is the equilibrium vapour pressure of component A in the vapour phase, $x_A$ is the mole fraction of the same component in the liquid phase and $P_A^0$ is the vapour pressure of the pure component, A, at the same temperature as the solution.

It follows, therefore, from the definition of mole fraction in the vapour phase, $y_A$, that

$$y_A = \frac{P_A}{P_T} = \frac{x_A P_A^0}{P_T} \qquad (4.47)$$

where $P_T$ is the total pressure in the vapour phase. For a binary system of components A and B,

$$\frac{y_A}{y_B} = \frac{P_A}{P_B} = \frac{x_A P_A^0}{x_B P_B^0} \qquad (4.48)$$

Equation (4.48) shows that the mole ratio of components in the vapour phase is not the same as their mole ratio in the liquid phase.

Raoult's law is found to hold, generally, for non-polar gases and vapours in non-polar liquids. It is useful in predicting the equilibrium composition of vapour/liquid mixtures of such systems.

For non-polar vapours in polar liquids, the vapour pressure of solution, P, is given by

$$P = \alpha k P_S \qquad (4.49)$$

where $\alpha$ is a correction factor, usually, determined experimentally, $k$ is a factor, dependent on concentration and $P_S$ is the vapour pressure of the pure solvent.

## Illustrative Example 4.13

Determine the composition of vapour in equilibrium with a liquid consisting of 50 % benzene in toluene. The system is at 70 C and you are given that the pure component vapour pressures of benzene and toluene, at this temperature, are 548.08 mm Hg and 202.62 mm Hg, respectively,

**Answer**

Let A represent benzene and B, toluene.

By Raoult's law, $P_A = x_A P_A^0$

The partial pressures are obtained as follows:

For benzene, $P_A = x_A P_A^0 = 0.5 \times 548.08 = 274.04 \ mm \ Hg$

For toluene, $P_B = x_B P_B^0 = 0.5 \times 202.62 = 101.31 \ mm \ Hg$

Hence, the total vapour pressure of the system = 274.04 + 101.31 = 375.35 mm Hg.

Mole fraction of benzene in the vapour phase = 274.04/375.35 = 0.73

Mole fraction of toluene in the vapour phase = 101.31/375,35 = 0.27

Ans.

**Illustrative Example 4.14**

Given the vapour pressure versus liquid composition data in the Tables below, for the benzene/toluene and chloroform/acetone systems, construct vapour pressure/vapour composition diagrams for the ideal benzene/toluene system and compare them with those for the non-ideal chloroform/acetone system.

**Table 1:  Vapour Pressure Data for the Benzene/ Toluene System**

| Mole % Benzene in vapour | Partial Pressure of Benzene, mm Hg | Partial Pressure of Toluene, mm Hg | Total Pressure, mm Hg |
|---|---|---|---|
| 0 | 0 | 202.62 | 202.62 |
| 20 | 109.62 | 162.10 | 271.72 |
| 40 | 219.23 | 121.57 | 340.8 |
| 60 | 328.85 | 81.05 | 409.9 |
| 80 | 438.46 | 40.52 | 478.98 |
| 100 | 548.08 | 0.00 | 548.08 |

**Table 2:  Vapour Pressure Data for the Chloroform/ Acetone System**

| Mole % Chloroform in vapour | Partial Pressure of Chloroform, mm Hg | Partial Pressure of Acetone, mm Hg | Total Pressure, mm Hg |
|---|---|---|---|
| 0 | 0 | 419.31 | 419.31 |

| 20 | 70.70 | 335.45 | 406.15 |
| 40 | 141.41 | 251.59 | 393.00 |
| 60 | 212.11 | 167.72 | 379.83 |
| 80 | 282.82 | 83.86 | 366.68 |
| 100 | 353.52 | 0.00 | 353.52 |

## Answer

Assuming that Raoult's law applies to the two systems, the mole fractions in the vapour and liquid phases are calculated from equations (4.47) and (4.46) and tabulated as shown in Tables 3 and 4

**Table 3:** Mole fractions in the Vapour and Liquid Phases for the Benzene/ Toluene System

| Mole % Benzene in vapour | Mole fraction Benzene in vapour, $y_A = P_A/P_T$ | Mole fraction Toluene in vapour, $y_B = 1 - y_A$ | Mole fraction Benzene in liquid, $x_A = P_A/P_A^o$ | Mole fraction Toluene in liquid, $x_B = 1 - x_A$ |
|---|---|---|---|---|
| 0 | 0 | 1.000 | 0 | 1.000 |
| 20 | 0.403 | 0.597 | 0.200 | 0.800 |
| 40 | 0.643 | 0.357 | 0.400 | 0.600 |
| 60 | 0.802 | 0.198 | 0.600 | 0.400 |
| 80 | 0.915 | 0.085 | 0.800 | 0.200 |
| 100 | 1.000 | 0 | 1.000 | 0 |

**Table 4:** Mole fractions in the Vapour and Liquid Phases for the Chloroform/ Acetone System

| Mole % Chloroform in vapour | Mole fraction Chloroform in vapour, $y_A = P_A/P_T$ | Mole fraction Acetone in vapour, $y_B = 1 - y_A$ | Mole fraction Chloroform in liquid, $x_A = P_A/P_A^o$ | Mole fraction Acetone in liquid, $x_B = 1 - x_A$ |
|---|---|---|---|---|
| 0 | 0 | 1.000 | 0 | 1.000 |
| 20 | 0.174 | 0.826 | 0.200 | 0.800 |
| 40 | 0.360 | 0.640 | 0.400 | 0.600 |
| 60 | 0.558 | 0.442 | 0.600 | 0.400 |
| 80 | 0.771 | 0.229 | 0.800 | 0.200 |
| 100 | 1.000 | 0 | 1.000 | 0 |

These data are plotted in the figure below. It can be seen that the benzene/toluene system deviates, positively, from the ideal solution

133

predicted by Raoult's law while the chloroform/acetone system deviates, negatively, ftom it.

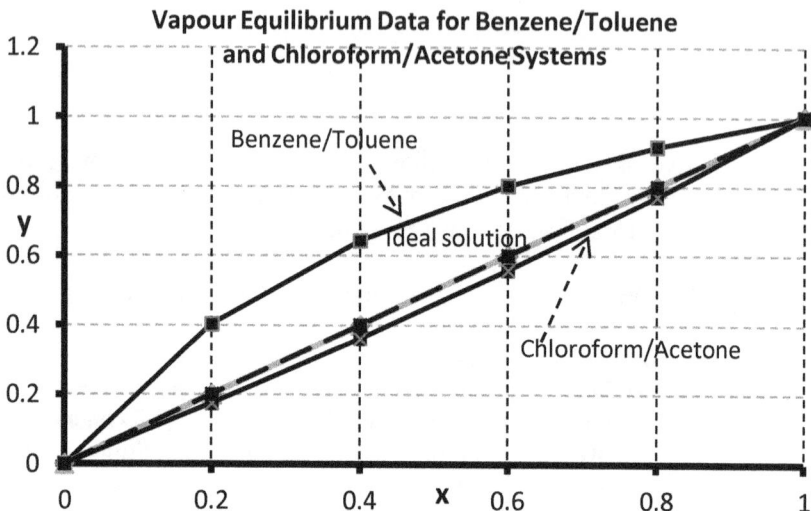

## 4.15.2: Non-Ideal Solutions

Solutions are non-ideal when their behaviour deviates from those of ideal solutions. One way of indicating this non-ideality is by the nature of the deviation from Rauolt's law.

For positive deviation from Raoult's law, the attraction between solute molecules in solution for each other exceeds the attraction between solute and solvent molecules. In vaporization, therefore, there is a greater tendency for them to leave the solution together. This means that their vapour pressure is higher than in an ideal solution. Thus, their vapour pressure versus molar concentration curve lies above that of an ideal solution.

For negative deviation from Raoult's law, the attraction between solute and solvent molecules in solution exceeds the attraction between solute molecules. In vaporization, therefore, there is a greater tendency for solute and solvent molecules to stay together and not leave the solution. This means that the vapour pressure of solute is lower than in an ideal solution. Thus, the vapour pressure versus molar concentration curve lies below that of an ideal solution.

These ideas may be illustrated diagramatically for various kinds of positive and negative deviation such as shown in curves A, B, and C in each of Figures 4.2 and 4.3. The dotted straight lines represent ideal solution comparison curves.

**Fig. 4.2: Equilibrium Composition Curves for Non Ideal Mixtures showing Positive Deviation from Ideality**

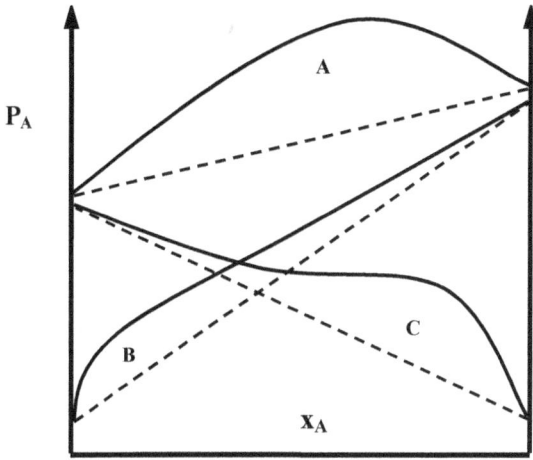

**Fig. 4.3: Equilibrium Composition Curves for Non Ideal Mixtures showing Negative Deviation from Ideality**

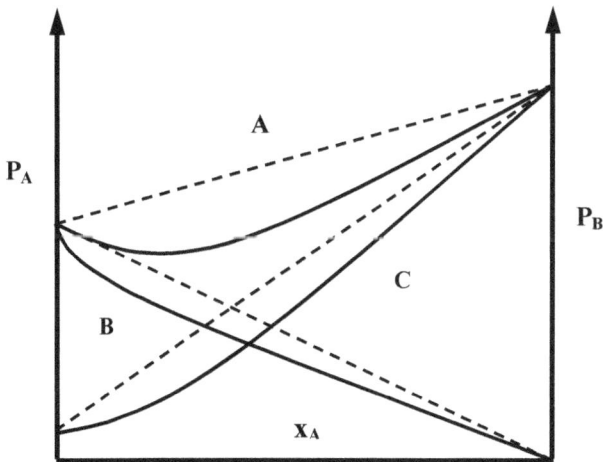

### 4.15.2.1: Vaporization Of Liquid Solutions Containing Non-Volatile Solutes

By Raoult's law, the total pressure, $P_T$, is given by

$$P_T = x_A P_A^0 + x_B P_B^0 = x_B P_B^0 \tag{4.50}$$

since $P_A^0. = 0$ for non-volatile solute.

The presence of non-volatile solute lowers the vapour pressure of solutions and raises the boiling point of solution compared to that of the solvent. The vapour pressure depression is proportional to the amount of non-volatile solute added. Elevation of boiling point is proportional to the amount of non-volatile solute added.

In industry, commercially important non-volatile solutes occur, frequently, as concentrated solutions which can only be removed by crystallization. In such cases, Raoult's and Henry's laws do not apply and a reference substance plot, the Duehring plot is used.

The Duehring plot is a reference substance plot which correlates the boiling points of solutions with that of water at the same pressure. When $P = P_W$,

$$\frac{d \ln P}{d \ln P_W} = \frac{-\dfrac{\Delta H}{R} d\left(\dfrac{1}{T}\right)}{-\dfrac{\Delta H_W}{R} d\left(\dfrac{1}{T_W}\right)} = 1 \tag{4.51}$$

or $\quad d(\dfrac{1}{T_W}) = \dfrac{\Delta H}{\Delta H_W} d(\dfrac{1}{T}) \quad$ that is $\quad \dfrac{1}{T_W} = \dfrac{\Delta H}{\Delta H_W} \cdot \dfrac{1}{T} + C \tag{4.52}$

If Trouton's rule holds, $C = 0$ and

$$T_W = \frac{\Delta H_W}{\Delta H} T \quad or \quad t_W = \frac{\Delta H_W}{\Delta H} t + C_1 \tag{4.53}$$

When $T = Tw$,

$$\frac{d \ln P}{d \ln P_W} = \frac{\Delta H}{\Delta H_W} \tag{4.54}$$

That is, on rearrangement and integration,

$$\ln P = \frac{\Delta H_V}{\Delta H_{V,W}} \ln P_W + cons \tan t \tag{4.55}$$

A graphical plot of $\ln P$ versus $\ln P_W$ when $T = T_W$ is known as the Cox chart. It looks like the Duerhing chart except that the axes are in logarithms of vapour pressures rather than in the inverse of the absolute temperatures as in the Duerhing chart.

**Fig. 4.4:  Form of the Duerhing Plot**

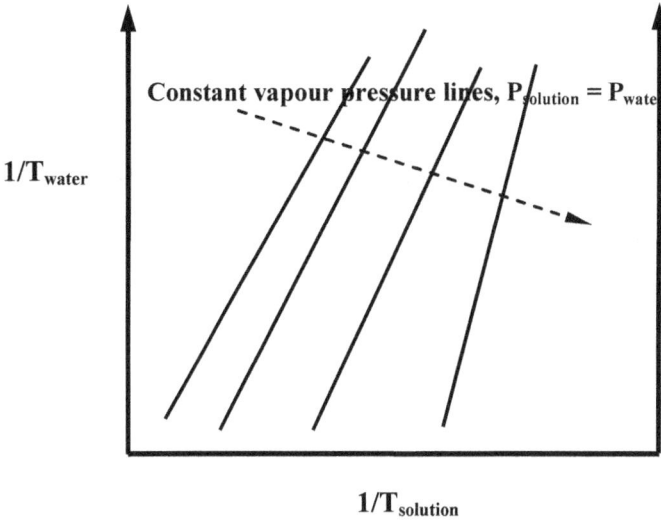

$1/T_{solution}$

The Duehring chart enables the experimental determination of the elevation of boiling point while the Cox chart helps to determine the lowering of the vapour pressure. The applicable equation is $P = k\,P_S$ where $P_S$ is the vapour pressure of pure solvent and $k$ is a factor dependent on concentration. For an ideal solution, $k$ = mole fraction of Raoult's law. The Duehring plot is an equal vapour pressure reference substance plot while the Cox plot is an equal -temperature reference substance plot.

### 4.15.2.2:  Vapour – Carrier Gas Systems

This situation is encountered in solvent recovery processes, in air conditioning, and in drying and humidification. A permanent, non-condensable gas, as the carrier gas, is used to transfer, transport or move condensable gas in and out of a system.

Bone dry gas is defined as gas absolutely devoid of condensable vapour, while saturation, at any temperature of the system, is defined by.

$$Relative\ saturation = \frac{vapour\ pressure\ of\ condensable\ gas}{its\ satd\ vapour\ pressure\ at\ same\ temp} = \frac{P_A}{P_A^o} \quad (4.56)$$

When the condensable vapour is water, humidity replaces saturation.

### 4.15.2.3: Solubility of Non - Condensable Gases - Henry's Law

While Raoult's law is useful for determining the partial pressure of a volatile (condensable) solute dissolved in a solution, Henry's law enables the solubility of a non-condensable gas in a solvent, to be determined.

Henry's law states that the equilibrium mole fraction of a gas, dissolved in a liquid, is, directly, proportional to the pressure of the gas above the liquid. That is:

$$P_A = H_A x_A \tag{4.57}$$

where $H_A$ is the Henry's constant for the gas, A, in units of pressure per mole fraction and $x_A$ is the equilibrium mole fraction of the gas, A, in the liquid.

Equation (4.57) is true only for dilute solutions. $H_A$ is, typically, large. It increases with temperature and is characteristic of a specific gas and liquid combination. Typical values are shown in Figure 4.5 below.

When a, sparingly, soluble gas is dissolved in a volatile solvent, Henry's law is, also, useful for estimating the solubility of the gas in the given solvent while Raoult's law is, still, used to estimate the partial pressure of the solvent in the gas-vapour mixture above the solvent.

### 4.15.2.4: Psychrometry–Vapour Pressures and Compositions of Mixtures of Condensable Vapour in Non-Condensable Gases

Humidity

The concept of humidity is used to describe solubility issues in systems in which condensable vapour is in a mixture with non-condensable gas. Examples of such systems include mixtures of condensable vapours such as water, benzene, ethanol, petrol, etc. with non-condensable gases such as oxygen, nitrogen, air, etc. Such systems, often, arise during the drying of materials, air conditioning of spaces, humidification and dehumidification of gaseous streams, and the purging of spaces and enclosures.

The terms and definitions, commonly used in humidity studies, also known as psychrometry, are outlined below. The air - water system is the most, frequently, encountered system. The definitions listed here, though based on water and air, apply to all systems as long as water and air are replaced by the condensable and non-condensable components in any other system under consideration. At any temperature:

$$Molal\ Humidity, Y_M = \frac{Moles\ of\ water\ vapour\ in\ the\ air - gas\ mixture}{Moles\ of\ dry\ gas\ in\ the\ mixture} \qquad (4.58)$$

$$= \frac{partial\ pressure\ of\ water\ vapour\ in\ the\ air - gas\ mixture}{partial\ pressure\ of\ dry\ gas\ in\ the\ mixture} \qquad (4.59)$$

Thus, if $P_W$ is the partial pressure of water vapour and $P_T$ is the total pressure in the system, then

$$Y_M = \frac{P_W}{P_T - P_W} \qquad (4.60)$$

$$Specific\ Humidity, Y = \frac{mass\ of\ water\ vapour\ in\ the\ air - gas\ mixture}{mass\ of\ dry\ gas\ in\ the\ mixture} \qquad (4.61)$$

$$= \frac{P_W}{P_T - P_W} \cdot \frac{M_W}{M_G} = Y_M \cdot \frac{M_W}{M_G} \qquad (4.62)$$

where $M_W$ and $M_G$ are the molecular weights of water and non-condensable gas, respectively.

$$Molal\ Humidity\ at\ saturation, Y_S = \frac{P_W^o}{P_T - P_W^o} \qquad (4.63)$$

where $P_W^o$ is the saturation vapour pressure of water at the temperature of the mixture.

$$Percent\ Humidity, Y_P = \frac{100\ Y_M}{Y_S} \qquad (4.64)$$

$$Percent\ relative\ Humidity, \%\ RH, Y_{RH} = \frac{100\ P_W}{P_W^o} \qquad (4.65)$$

It can be seen from (4.60), (4.64) and (4.65) that

$$Percent\ Humidity, Y_P - Y_{RH} \frac{P_T - P_W^o}{P_T - P_W} \qquad (4.66)$$

**Illustrative Example 4.15**

The weather condition in Enugu, Nigeria, on a normal sunny day, is such that the air temperature is 29 C, the atmospheric pressure is 758 mm. Hg and the relative humidity is 70 %. Determine the humidity of Enugu air in terms of the various definitions of humidity above.

**Fig 4.5: Henry's Law Constant as a Function of Temperature (Schmidt & List, 1962)**

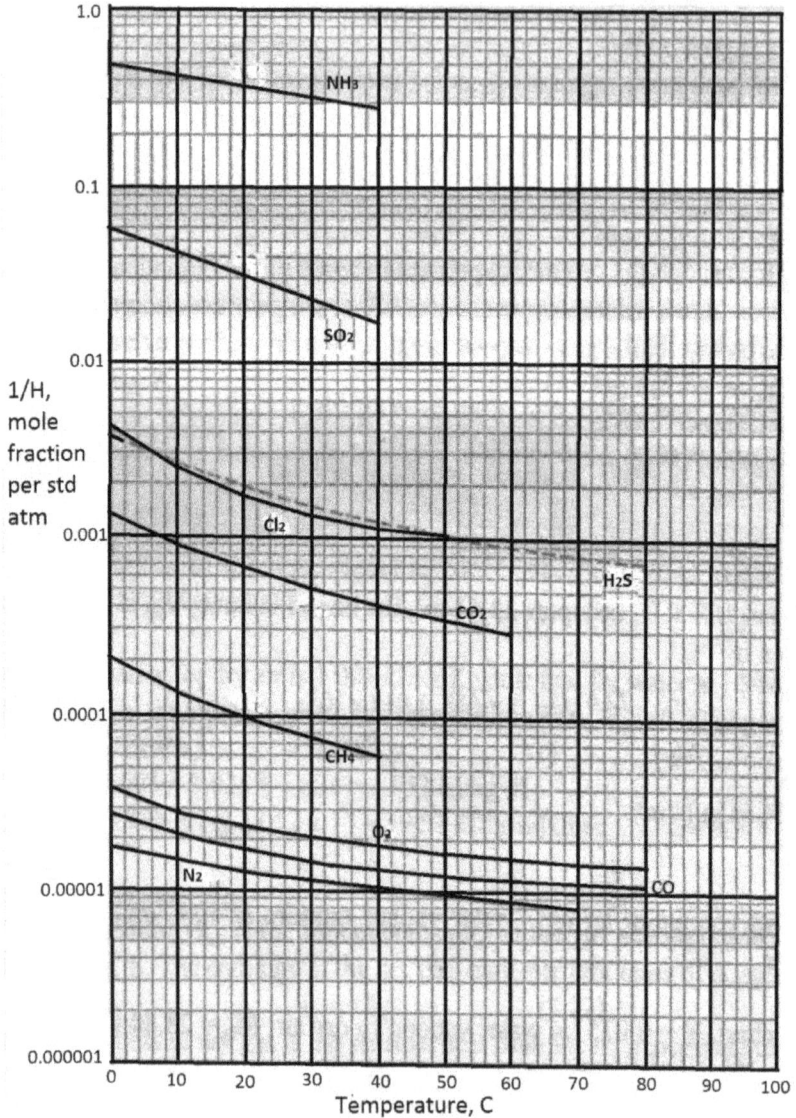

**Answer**

The saturation vapour pressure of water at 29 C, $P_w{}^\circ = 29.6$ mm. Hg. Since the relative humidity is 70%,

$$Y_{RH} = 70 = \frac{100 \, P_W}{P_W^o} = \frac{100 \, P_W}{29.6} \quad \text{or} \quad P_W = \frac{70 \, x \, 29.6}{100} = 20.72 \, mm \, Hg \tag{1}$$

From (4.60)

$$Molar \; humidity = Y_M = \frac{P_W}{P_T - P_W} = \frac{20.72}{758 - 20.72} = 0.028 \, \frac{moles \; water \; vapour}{mole \; dry \; air} \tag{2}$$

Before we can compute the specific humidity of the air, we need to know the molecular weight of dry air, which is a mixture of oxygen and nitrogen. Since dry air is 79 % nitrogen and 21 % oxygen, its molecular weight, $M_G$, is given by

$$M_G = 0.79 \, x \, 28 + 0.21 \, x \, 32 = 28.84 \tag{3}$$

The molecular weight of water vapour, $M_W$, is 18. Hence, from (4.62), the specific humidity, Y, is

$$Y = Y_M \cdot \frac{M_W}{M_G} = 0.028 \, x \, \frac{18}{28.84} = 0.018 \, \frac{kg \; water \; vapour}{kg \; dry \; air} \tag{4}$$

From (4.63), the molal humidity at saturation, $Y_S$, is

$$Y_S = \frac{P_W^o}{P_T - P_W^o} = \frac{29.6}{758 - 29.6} = 0.041 \, \frac{moles \; water \; vapour}{mole \; dry \; air} \tag{5}$$

From (4.64), the percent humidity, $Y_P$, is

$$Percent \; Humidity, \; Y_P = \frac{100 \, Y_M}{Y_S} = \frac{100 \, x \, 0.028}{0.041} = 68.3\% \tag{6}$$

Note that this is different from percent relative humidity.

Humidity Charts

These are diagrams which show the relationships between humidity and temperature of a gas - vapour mixture. Fig. 4.6 shows a typical humidity chart for the air - water system. The important relationships shown in complete versions of this chart are:

i) Humidity versus dry gas temperature curves at various percent relative humidity
ii). Lines of constant humidity at isothermal and adiabatic saturation
iii). Lines showing the latent heat of vaporisation versus dry gas temperature
iv). Enthalpy per unit mass of gas - vapour mixture (humid heat) versus humidity
v). Specific volumes of dry air as a function of temperature and humidity.

To use this chart effectively, one needs to understand the following terms associated with the chart.

**Fig. 4.6: A Psychromteric (Humidity) Chart for Air (Hanna R., 2010)**

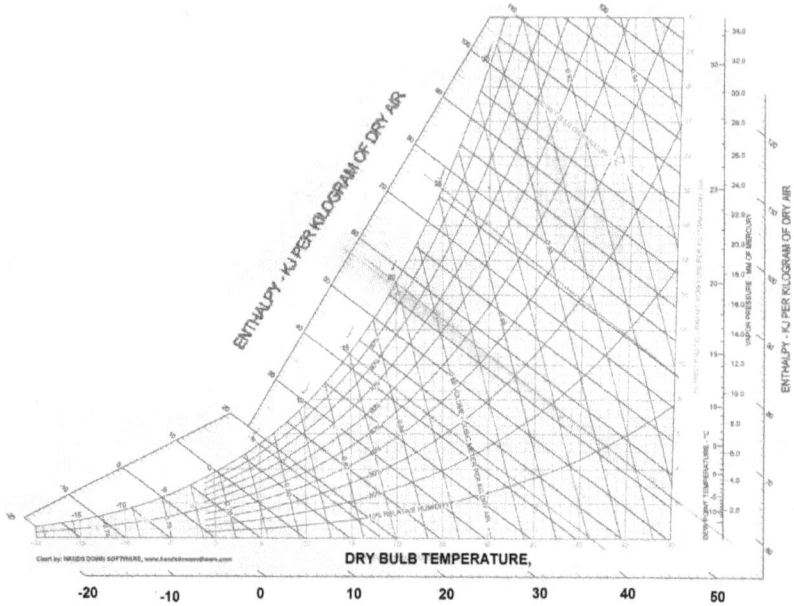

## Dry Bulb Temperature

This is the actual temperature of the vapour/gas mixture as measured by a thermometer or any suitable temperature measurement device inserted in the mixture.

## Dew Point

This is the temperature at which the vapour/gas mixture just becomes saturated as a result of being cooled but in which the cooling is not because the vapour/gas mixture is in contact with a liquid. A good illustration of the dew point can be seen when a cold bottle of beer is first taken out of the refrigerator. It cools the air surrounding it to a temperature at which the water vapour in the surrounding air begins to condense as a liquid on the outside of the beer bottle. This temperature, at which condensation just begins, is the dew point of the surrounding air.

142

## Humid Volume

This is the volume of the vapour/gas mixture per unit mass of dry (vapour free) gas at the prevailing temperature and pressure. We can calculate the mass in moles of the gas/vapour mixture as

$$\frac{Moles\ of\ gas-vapour\ mixture}{mass\ dry\ gas} = \frac{moles\ dry\ gas}{mass\ dry\ gas} + \frac{moles\ of\ vapour}{mass\ dry\ gas} \quad (4.67)$$

$$= \frac{mass\ dry\ gas}{M_G\left(mass\ dry\ gas\right)} + \frac{mass\ of\ vapour}{M_V\left(mass\ dry\ gas\right)}$$

$$= \frac{1}{M_G} + \frac{Y}{M_V} \quad (4.68)$$

where $M_G$ is the molecular weight of the dry gas, $M_V$ that of the vapour and Y the specific humidity of the mixture. If we assume that the gas/vapour mixture is an ideal gas, we can assume that at S. T. P., one kmol of it will occupy 22.41 m$^3$.

We can derive from the ideal gas law states that

$$\frac{P_1 V_1}{P_2 V_2} = \frac{n_1\ R T_1}{n_2\ R T_2} \quad or \quad V_2 = \frac{n_2\ T_2}{n_1\ T_1} \frac{P_1 V_1}{P_2} \quad (4.69)$$

If we the denote humid volume at temperature, $T_2$, and pressure, $P_2$, by $V_H$ and take $T_1$, $P_1$ and $V_1$ to be S. T. P conditions, then equation (4.69) becomes

$$V_H = \left(\frac{1}{M_G} + \frac{Y}{M_V}\right) \frac{\left(t_G + 273\right)}{273} \frac{1.013 \times 10^5 \times 22.41}{P_2}, \frac{moles\ mixture}{mass\ dry\ gas} \cdot \frac{m^3 mixture}{moles\ mixture}$$

$$= 8315.5 \left(\frac{1}{M_G} + \frac{Y}{M_V}\right) \frac{\left(t_G + 273\right)}{P_2}, \frac{m^3\ mixture}{mass\ dry\ gas} \quad (4.70)$$

## Humid Heat

This is the heat energy required, at constant pressure, to raise the temperature of a unit mass of gas plus its associated vapour one degree. That is

$$C_S = Cp_G + Y Cp_V \quad (4.71)$$

where Cs is the humid heat capacity of the mixture, kJ/kg.K; $Cp_G$ is the heat capacity of dry gas, kJ/kg.K; $Cp_V$ that of the vapour and Y, the absolute humidity of the mixture, kg vapour/kg dry gas.

## Enthalpy

The enthalpy of the vapour/gas mixture is the sum of the enthalpies of the vapour and the gas at the temperature of the mixture, $t_G$, referred to the temperature of a standardized reference state, $t_o$.

The enthalpy of the dry gas alone, $H_G$, per unit mass of dry gas, is given by

$$H_G = Cp_G (t_G - t_o) \qquad (4.72)$$

The enthalpy of the vapour alone, $H_V$, per unit mass of dry gas, is given by

$$H_V = Cp_L (t_{DP} - t_o) + \lambda_{DP} + Cp_V (t_G - t_{DP}) \qquad (4.73)$$

where $Cp_L$ is the heat capacity of the vapour when it is in the liquid state, kJ/kg.K; $t_{DP}$ is the dew point temperature, K and $\lambda_{DP}$ is the latent heat of vaporization at the dew point, kJ/kg vapour.

At low total pressures, such as under 20 atm, the enthalpy of the vapour can be approximated, without much loss inn accuracy, by

$$H_V = \lambda_o + Cp_V (t_G - t_o) \qquad (4.74)$$

where $\lambda_o$ is the latent heat of vaporization, kJ/kg vapour, at the standardized reference state temperature, $t_o$.

Thus, the enthalpy, $H_M$, of the vapour/gas mixture, is

$$H_M = H_G + Y H_V = Cp_G (t_G - t_o) + Y [\lambda_o + Cp_V (t_G - t_o)]$$
$$= (Cp_G + YCp_V)(t_G - t_o) + Y \lambda_o = C_S (t_G - t_o) + Y \lambda_o \qquad (4.75)$$

## Wet Bulb Temperature

This is the steady state temperature attained by a small amount of liquid evaporating in a large environment of unsaturated gas-vapour mixture. A typical system is that of a water droplet falling in air or a wet wick surrounding the bulb of a mercury-in-glass thernometer.

As a result of evaporation, the temperature of the bulb or wick continues to fall until it reaches a steady state value called the wet bulb temperature. A heat and mass balance on the drop of liquid, evaporating in the unsaturated gas-vapour mixture, gives the rate of heat loss from the drop, $R_G$, as

$$R_G = h_G (t_G - t_W) . A_D \qquad (4.76)$$

where $h_G$ is the heat transfer coefficient in the neighborhood of the drop, kW/m².K, $t_w$ is the wet bulb temperature, K, and $A_D$ is the surface area of the drop, $m^2$.

The rate at which heat is gained by the vapour leaving the drop during
evaporation, $R_V$, is

$$R_V = N_V \lambda_W = k_Y \left(Y_W - Y\right).A_D.\lambda_W \qquad (4.77)$$

where $N_V$ = rate of vapour loss from the drop, $kg/s.m^2$, $\lambda_W$ is the latent
heat of vaporization at the wet bulb temperature, kJ/kg vapour, $k_Y$ is the
mass transfer coefficient; $Y_W$ is the absolute humidity in the
neighborhood of the drop and $Y$ is the absolute humidity of the
environment.

At steady state, the heat lost by the drop during evaporation is equal to
the heat carried into the environment of the vapour/gas stream by the
evaporating vapour. That is

$$R_G = R_V \quad or \quad h_G \left(t_G - t_W\right).A_D = k_Y \left(Y_W - Y\right).A_D.\lambda_W$$

Thus

$$t_G - t_W = \frac{\left(Y_W - Y\right).\lambda_W}{h_G/k_Y} \qquad (4.78)$$

$(t_G - t_W)$ is known as the wet bulb depression and is, usually, measured
with the wet and dry bulb thermometers. $h_G/k_Y$ is known as the
psychrometric ratio. These two parameters form the basis for the
determination of the humidity of a vapour/gas mixture environment.
$h_G/k_Y$ is often estimated from dimensionless correlations of which the
more popular is that quotyed by Treybal (1980) as

$$\frac{h_G}{k_Y\,C_S} = \left(\frac{Sc}{Pr}\right)^{0.567} = Le^{0.567} \qquad (4.79)$$

where Cs is the humid heat, Sc the Schmidt number, Pr the Prandtl
number and Le, the Lewis number. For most systems

$$\frac{h_G}{k_Y} = 950 \frac{N\,m}{kg\,dry\,gas.K} \qquad (4.80)$$

For the air-water system, the Lewis number is equal to 1 so that, from
equation (4.79)

$$\frac{h_G}{k_Y} = C_S \qquad (4.81)$$

Equation (4.81) is known as the Lewis relation.

The illustrative examples, given below, demonstrate the more common
calculations and parameter usage in respect of the humidity chart.

## Illustrative Example 4.16

A weather report states that the temperature of air is 30 C with a relative humidity of 85 % and a barometric pressure of 739 mm Hg. Calculate

a)      the molar humidity of air
b)      the humidity
c)      the % humidity
d)      the saturation temperature or dew point of the air
e)      the degrees of superheat of vapour
f)      if the air is heated to 41 C at the same pressure, determine the molal humidity and dew point of the air

## Answer

a)   Molar humidity of air

From steam tables, at 30 C, $P_W^o = 0.04242$ bar.
From relative humidity equation,

$$Y_R = \frac{p_W}{p_W^0} x100 = 85 \quad or \quad p_W = 0.85 \, x \, 0.04242 = 0.03606 \, bar \tag{1}$$

$$P_T = \frac{739}{750} = 0.9853 \, bar \tag{2}$$

From the molar humidity equation,

$$Y_M = \frac{p_W}{p_G} = \frac{p_W}{P_T - p_W} = \frac{0.03606}{0.9853 - 0.03606} = 0.038 \qquad Ans.$$

b)   Absolute humidity of air

Air is 21 % oxygen and 79 % nitrogen
$$M_{air} = 0.21 \, x \, 32 + 0.79 \, x \, 28 = 28.84 \tag{3}$$
From the humidity equation
$$Y = \frac{p_W M_W}{p_G M_G} = Y_M \frac{M_W}{M_G} = 0.038 \, x \frac{18}{28.84} = 0.024 \qquad Ans.$$

c)   Percent Humidity of air

146

Humidity at saturation

$$Y_S = \frac{p_W^o}{p_G} = \frac{p_W^o}{P_T - p_W^o} = \frac{0.04242}{0.9853 - 0.04242} = 0.045 \qquad (4)$$

$$\% \text{ Humidity } = Y_P = \frac{Y_M}{Y_S} x 100 = \frac{0.038}{0.045} x 100 = 84.4\% \qquad Ans.$$

a)      Saturation temperature of air

At saturation, $P_W = P_W^o = 0.03606$ bar

From steam tables, this corresponds to

$$t_W = 27 + 1 x \frac{(0.03606 - 0.03564)}{(0.03778 - 0.03564)} = 27.2\,C, \text{ the dew } po \text{ int} \qquad Ans.$$

b)      Degrees of super heat  $=$  30 - 27.2 $= 2.8$ C          Ans.

c)      Molal humidity and dew point of heated air

      $Y_M$ and $t_W$ are not changed by heating          Ans.

**Illustrative Example 4.17**

Hot air at 40 C and 70 % relative humidity is to be cooled to 25 C and to a relative humidity of 50 %. Determine how much water vapour will have to be removed from the hot air per unit mass of dry air. What will be the final humidity of the air?

**Answer**

At 40 C and 70 % relative humidity, the absolute humidity of air is read from the chart to be 0.035 kg water vapour per kg of dry air (obtained by reading the point of intersection on the absolute humiditry axis, of the vertical line on 40 C from the temperature axis and the 70 % relative humidity curved line).

At 25 C and 50 % relative humidity, the absolute humidity of air is, similarly, read from the chart to be 0.01 kg water vapour per kg of dry air. The amount of water vapour to be removed is, thus, 0.035 − 0.01 = 0.025 kg water vapour per kg dry air. Ans.

## Illustrative Example 4.18

Calculate the humid volume of an air-water vapour mixture at 40 C, 70 % relative humidity and 1 atm pressure.

## Answer

From the humidity chart, $Y = 0.035$ kg water vapour per kg dry air. The molecular weights of dry air and water vapour are, respectively, $M_G = 28.84$ and $M_V = 18$. $t_G = 40$ C and $P_T = 1.013 \times 10^5$ N/m$^2$. Thus, from equation (4.68)

$$V_H = 8315.5 \left( \frac{1}{M_G} + \frac{Y}{M_V} \right) \frac{(t_G + 273)}{P_2}, \frac{m^3 \ mixture}{mass \ dry \ gas}$$

$$= 8315.5 \left( \frac{1}{28.84} + \frac{0.035}{18} \right) \frac{(40 + 273)}{1.013 \times 10^5}, \frac{m^3 \ mixture}{mass \ dry \ gas} = 0.94 \frac{m^3}{kg} \quad Ans$$

## Illustrative Example 4.19

Calculate the heat energy required to heat an air - water vapour mixture from 30 C to 40 C. The relative humidity of the air - water vapour mixture at 30 C was 90 %.

## Answer

From standard tables, the heat capacity for dry air, in the temperature range of 30 to 40 C is 1.005 kJ/kg.K while that for water is 1.884 kJ/kg.K.

From the humidity chart, the absolute humidity for air, Y, at 30 C and 90 % relative humidity, is 0.025 kg water vapour/kg dry air. At 40 C, the relative humidity will be 50 % even though the absolute humidity of the air would still be constant at 0.025 kg water vapour /kg dry air.

Hence the humid heat would, also, be constant in this temperature range and would be given by equation (4.71) as

$$C_S = Cp_G + Y Cp_V = 1.005 + 0.025 \times 1.884 = 1.052 \, kJ/kg.K \quad (1)$$

Heat energy required, Q, is, therefore,

$$Q = C_S (t_2 - t_1) = 1.052 (40 - 30) = 10.52 \, kJ/kg \ mixture \quad Ans.$$

## Illustrative Example 4.20

Calculate the enthalpy of an air-water vapour mixture at 50 C and 40 %
relative humidity.

### Answer

From the psychrometric chart, the absolute humidity of this system is
0.035 kg water vapour/kg dry air.
$$C_S = Cp_G + YCp_V = 1.005 + 0.035 \times 1.884 = 1.071 \, kJ/kg.K \tag{1}$$
Since $t_o = 0$ and $\lambda_o = 2502.3$ kJ/kg, equation (4.75) gives
$$H_M = C_S(t_G - t_o) + Y\lambda_o = 1.071 \times 50 + 0.035 \times 2502.3 = 141.13 \, kJ/kg \quad Ans$$

## Illustrative Example 4.21

A wet and dry bulb thermometer, in a room at 32 C and 1 atm, gave a
wet bulb depression of 10 C. Estimate absolute humidity of the room.

### Answer

Since $t_G - t_W = 10$ C, then $t_W = 32$ C - 10 C = 22 C $\hspace{2cm}$ (1)

From steam tables, at 32 C,
$\lambda_W = h_V - h_L = 2541 - 91.6 = 2449.4$ kJ/kg and $P_W = 0.02617$ bar $\hspace{0.5cm}$ (2)
From equation (4.61) the absolute humidity, $Y_W$, is
$$Y_W = \frac{P_W}{P_T - P_W} \cdot \frac{M_W}{M_G} = \frac{0.02617}{1.013 - 0.02617} \cdot \frac{18}{28.84} = 0.0166 \frac{kg \ water \ vapour}{kg \ dry \ air} \tag{3}$$

Thus, from equation (4.78)
$$t_G - t_W = \frac{(Y_W - Y).\lambda_W}{h_G/k_Y} \quad or \quad Y = Y_W - \frac{(t_G - t_W).h_G/k_Y}{\lambda_W} \tag{4}$$

From equation (4.80)
$$\frac{h_G}{k_Y} = 950 \frac{N \, m}{kg \ dry \ air.K} \tag{5}$$

so that
$$Y = Y_W - \frac{(t_G - t_W).h_G/k_Y}{\lambda_W}$$

$$= 0.0166 - \frac{0.950 \times 10 \ K}{2449.4} \cdot \frac{kNm}{kg \ dry \ air.K} \cdot \frac{kg \ vapour}{kJ} = 0.0127 \frac{kg \ vapour}{kg \ dry \ air} \ Ans.$$

149

### 4.15.2.5: *Vaporization Of Immiscible Liquids*

When a mixture of immiscuible liquids is heated, the mixture will boil when the sum of its pure component vapour pressures equals atmospheric pressure. The pure component vapour pressures of the component liquids can be related as

$$\frac{P_A^0}{P_B^0} = \frac{y_A}{y_B} = \frac{m_A}{M_A} \cdot \frac{M_B}{m_B} \tag{4.82}$$

The total pressure of the system is still the sum of the partial pressures of the components but the boiling point of the resulting mixture is lower than that of either of the components, a fact used to advantage in separating heat sensitive materials in such mixtures.

The most common application of this principle is in steam distillation where steam, as the carrier gas, is used to effect the extraction of a desired component of such a mixture. The boiling point of steam is, usually, arranged to be less than the temperature at which the other liquid may undergo thermal damage (heat sensitive materials can, also, be distilled under reduced pressure).

### Illustrative Example 4.22

Dodecane is insoluble in water and is to be purified by steam distillation at 750 mm Hg pressure.

d)    At what temperature will the distillation occur?
e)    How many degrees below the normal boiling point of dodecane is this?
f)    How many kg of steam will be, theoretically, required to distil one kg of dodecane

### Answer

From a Cox chart, the normal boiling point. of dodecane is estimated to be 217 C (accurate value is 216.3 C) while that of water is estimated as 100 C.

Since the boiling point of mixture is less than that of either component, the procedure is to guess a temperature below these, find, separately, the pure component vapour pressures of each component, and add them

up. If the sum is 750 mm Hg, you have got the answer. If not, guess another value until you get the sum equaling 750 mm Hg.
Hence for <u>Part (a)</u>

Guess t = 99 C. This gives $P_W = 733$ mm Hg and $P_{Dodecane} = 17$ mm Hg. The total vapour pressure of the system is then
$$P_T = P_W + P_{Dodecane} = 733 + 17 = 750 \tag{1}$$
Hence mixture boils at 99 C. Ans

<u>For Part (b)</u>

The reduction in boiling point is 217 - 99 = 118 C. Ans

<u>For Part (c)</u>

The molecular weight of dodecane ($C_{12}H_{26}$) is 170 while that of water ($H_2O$) is 18.
From equation (4.82), amount of steam, $m_W$, required to distil one kg of dodecane is
$$m_W = \frac{P_W M_W}{P_A M_A} \times m_A = \frac{733 \times 18}{17 \times 170} \times 1.0 = 4.57 \text{ kg steam. Ans}$$
Alternatively, if a Cox chart is not available or greater accuracy is required, the pure component vapour pressures can be estimated from the Antoine equations. These are for dodecane and water
$$Dodecane : \ln P_A = 16.1134 - \frac{3774.56}{T - 91.31}; \quad Water : \ln P_W = 18.3036 - \frac{3816.44}{T - 46.13} \tag{2}$$
which at t = 99 C or T = 99 + 273 = 372 K, give, for dodecane
$$\ln P_A = 16.1134 - \frac{3774.56}{372 - 91.31} = 2.6660 \quad or \quad P_A = 14.38 \text{ mm Hg} \tag{3}$$
and for water
$$\ln P_W = 18.3036 - \frac{3816.44}{372 - 46.13} = 6.5920 \quad or \quad P_W = 729.24 \text{ mm Hg} \tag{4}$$
So that, for Part (a)
$$P_T = P_W + P_{Dodecane} = 729.24 + 14.38 = 743.62 \tag{5}$$
This is below the 750 mm Hg specified in the problem. Although the deviation is less than 1% and, therefore, negligible for industrial purposes, we shall try to get a more accurate value in order to illustrate the procedure.

Thus, another guess of t = 99.4 C or T = 99.4 + 273 = 372.4 K gives, for dodecane

$$\ln P_A = 16.1134 - \frac{3774.56}{372.4 - 91.31} = 2.6851 \quad or \quad P_A = 14.66 \; mm \; Hg \tag{6}$$

and for water

$$\ln P_W = 18.3036 - \frac{3816.44}{372.4 - 46.13} = 6.6064 \quad or \quad P_W = 739.81 \; mm \; Hg \tag{7}$$

So that, for Part (a)

$$P_T = P_W + P_{Dodecane} = 739.81 + 14.66 = 754.47 \tag{8}$$

Although the error is still less than 1%, let us guess t = 99.3 C or T = 99.3 + 273 = 372.3  K. This gives, for dodecane

$$\ln P_A = 16.1134 - \frac{3774.56}{372.3 - 91.31} = 2.6803 \quad or \quad P_A = 14.59 \; mm \; Hg \tag{9}$$

and for water

$$\ln P_W = 18.3036 - \frac{3816.44}{372.3 - 46.13} = 6.6028 \quad or \quad P_W = 737.16 \; mm \; Hg \tag{10}$$

So that, for Part (a)

$$P_T = P_W + P_{Dodecane} = 737.16 + 14.59 = 751.75 \tag{11}$$

Further trials may show that T lies between 372.2 and 372.3 K.

### 4.15.2.6:  Vaporization Of Immiscible Liquids With Superheated Steam (Superheated Steam Distillation)

Equation (4.82) gives the minimum amount of carrier gas required. When steam is the carrier gas, the amount needed per unit quantity of liquid distilled out (distillate) can be varied independently by raising the temperature or lowering the total pressure of the system. Superheating the steam is an effective way of doing so.

Since the partial pressure of immiscible liquid is its pure component vapour pressure, $P_A^0$, at a total pressure of $P_T$, the the partial pressure of steam, $P_S$, in the vapour mixture is

$$P_S = P_T - P_A^o \tag{4.83}$$

which, when substituted in equation (4.82), gives

$$\frac{m_S}{m_A} = \frac{P_S M_S}{P_A^0 M_A} = \frac{M_S \left( P_T - P_A^0 \right)}{P_A^0 M_A} \tag{4.84}$$

For some mixtures, the actual vapour pressure of component A is not its pure component vapour pressure, $P_A^0$, but some other pressure, $P_A$.

When this occurs, $P_A < P^0_A$ and equation (4.84) becomes

$$\frac{m_S}{m_A} = \frac{P_S M_S}{P_A M_A} = \frac{M_S (P_T - P_A)}{P_A M_A} \qquad (4.85)$$

In such cases, a *vaporization efficiency*, $\beta$, can be defined as

$$\beta = \frac{P_A}{P^0_A} \qquad (4.86)$$

on the basis of which

$$\frac{m_S}{m_A} = \frac{M_S (P_T - \beta P^0_A)}{\beta P^0_A M_A} \qquad (4.87)$$

Note that when $Pa = P^0_A$, $\beta = 1$.

**Illustrative Example 4.23**

Calculate the steam requirement for distilling one kg of dodecane at a total pressure of 380 mm Hg if the safe distillation temperature is 150 C. Assume 100 % vaporization efficiency.

**Answer**

From the Cox chart, the vapour pressure of dodecane at 150 C is 120 mm Hg. Putting this in equation (4.84) we get

$$m_S = \frac{M_S (P_T - P^0_A)}{P^0_A M_A} \times m_A$$

$$= \frac{18 x (380 - 120)}{120 x 170} x 1.0 = 0.23 \, kg \, steam. / kg \, dodecane \, Ans$$

Alternatively, the vapour pressure of dodecane may be determined using the Antoine equation.

### 4.15.3: Prediction of Heats of Transition of Pure Substances

Heat of Transition

This is the heat energy absorbed or evolved, at constant temperature and pressure, during a transition or equilibrium change from liquid to solid (freezing), liquid to vapour (vaporisation) or vapour to solid (sublimation) or vice versa. This heat, also, known as the latent heat, in older usage, and as heat energy of transition, in modern times, is defined as the heat energy change which occurs when one unit mass or mole of substance undergoes the given transition at standard, atmospheric, pressure. Several types of transitions are possible and are listed in Table 4.5 below.

**Table. 4.5: Common Phase Transitions**

| Type of Transition | Common Name | Reverse Process |
|---|---|---|
| solid to liquid | heat or enthalpy of fusion | heat or enthalpy of melting |
| liquid to vapour | heat or enthalpy of vaporization | heat or enthalpy of condensation |
| solid to vapour | heat or enthalpy of sublimation | heat or enthalpy of condensation |
| solid to solid | heat or enthalpy of transformation | heat or enthalpy of transformation |

Heats of transition, often, need to be estimated for purposes of design or analysis. It is always best to use experimentally determined values which are, generally, available for many substances in standard text or in reference handbooks. In very rare cases, when experimental values are not available, they may be estimated by means of the common methods listed below.

Estimating the Heat of Fusion, $\Delta H_f$

The heat of fusion may be estimated from industrial experience and entropy calculations (Hougen et al, 1954) which show that,

for solid elements,
$$\frac{\Delta H}{T_f} = 9 \ to \ 14 \qquad (4.88) ;$$

for inorganic solids,
$$\frac{\Delta H}{T_f} = 24 \ to \ 33 \qquad (4.89)$$

;

and for organic solids,
$$\frac{\Delta H}{T_f} = 54 \qquad (4.90)$$

where $\Delta H_f$ is the heat of fusion, J/mol, and $T_f$ is the temperature of fusion, K.

Estimating the Heat of Vaporisation, $\Delta H_V$

Several methods, of varying accuracy, are available for the estimation of the heat of vaporisation. Only the most commonly used five will be

154

listed.

## a). Trouton 's Rule

This states that, for liquids;

$$\frac{\Delta H}{T_B} = 100 \qquad (4.91)$$

$\Delta H_V$ is the heat of vaporisation, J/mol, and $T_B$, the normal boiling point, K. Trouton's rule gives only very approximate values of the latent heat of vaporization. For example, for water, a polar liquid with a boiling point of 100 C or 373 K, Trouton's rule gives $\Delta H_V = 373 \times 100 = 37,300$ J/mol while the actual value is 40683 J/mol. Thus the value, estimated with Trouton's rule, is -8.3 % in error. For n-hexane, a non-polar liquid with a boiling point of 68.7 C or 341.7 K, Trouton's rule gives $\Delta H_V = 341.7 \times 100 = 34,170$ J/mol. Experimental value is 28,872 J/mol, Thus, Trouton's rule estimate is about +18.3 % in error.

## b)    The Clausius - Clayperon Equation

This is the equation which describes the relationship between the vapour pressure and temperature of a pure substance. It is stated, generally, as

$$\frac{dP}{dT} = \frac{\Delta H}{T \Delta V} = \frac{\Delta H_V}{T(V_G - V_L)} \qquad (4.92)$$

where P and T are, respectively, the absolute pressure and temperature, $\Delta H_V$ is the molar heat of vaporization and $V_G$, $V_L$ are the molar volumes of vapour and liquid phases of the substance.

Assuming that $\Delta H_V$ is not a function of T and that $V_G > V_L$, for an ideal gas, $V_G = RT/P$, equation (4.92) can be re-arranged as

$$\frac{1}{P}\frac{dP}{dT} = \frac{\Delta H_V}{RT^2} \qquad or \qquad d\ln P = -\frac{\Delta H_V}{R} d\left(\frac{1}{T}\right) \qquad (4.93)$$

which, on integration, gives

$$\ln\left(\frac{P_2}{P_1}\right) = -\frac{\Delta H_V}{R}\left(\frac{1}{T_2} - \frac{1}{T_1}\right) \qquad (4.94)$$

Subscripts 1 and 2 represent initial and final conditions of absolute temperature and vapour pressure and R is the gas constant (8.314 J mol$^{-1}$K$^{-1}$). The latent heat of vaporization may be estimated from either equation (4.94) or directly from equation (4.92).

Estimation from equation (4.92) requires a knowledge of the P versus T curve for the substance for which $\Delta H_V$ needs to be determined. This knowledge is provided by the Antoine equation, plotted below as curve C in Fig. 4.7. A tangent, such as line AB, at a particular temperature such as 373 K, gives the slope, $dP/dT$ of equation (4.92) at that temperature, from which, knowing $V_G$ and $V_L$ the molar volume of vapour and liquid phase, respectively, at that temperature, $\Delta H_V$ can be evaluated.

In a more analytical approach, the Antoine equation can be differentiated with respect to T to give

$$\frac{dP}{dT} = \frac{PB}{(T+C)^2} \tag{4.95}$$

It can be seen from equations (4.92) and (4.95) that

$$\frac{dP}{dT} = \frac{\Delta H_V}{T(V_G - V_L)} = \frac{PB}{(T+C)^2}$$

From which we get that

$$\Delta H_V = \frac{PBT(V_G - V_L)}{(T+C)^2} \tag{4.96}$$

Since $V_G$ is very much larger than $V_L$, $(V_G - V_L)$ in equation (4.88) can be replaced by $V_G$ without much loss in accuracy.

For an ideal gas, $V_G = RT/P$ so that equation (4.92) becomes

$$\Delta H = \frac{RBT^2}{(T+C)^2} \tag{4.97}$$

c) Use of the Haggenmacher Equation

Assuming that $V_G$ is very much larger than $V_L$, so that $(V_G - V_L)$ in equations (4.92) and (4.96) can be replaced by $V_G$ without much loss in accuracy and also that a real gas may be approximated to an ideal gas by the applying its compressibility constant to the ideal gas equation, we can get that $PV_G = \Delta ZRT$ or $V_G = \Delta ZRT/P$, where $\Delta Z$ is the difference between the compressibility constants of the vapour and liquid phases.

When this value of $V_G$ is substituted in (4.96) and R is replaced by its numerical value, the Haggenmacher equation (Coulson, Richardson & Sinnott, 1983) is obtained as

$$\Delta H_V = \frac{8.32 B \Delta Z T^2}{(T + C)^2}$$  (4.98)

where B and C are still the coefficients of the Antoine equation, Z is the compressibility constant such that

$$\Delta Z = Z_{vapour} - Z_{liquid} = \left(1 - \frac{P_R}{T_R}\right)^{0.5}$$  (4.99)

and $P_R$ and $T_R$ are the reduced pressure and temperature, respectively.

**Fig. 4.7:  P vs T Curve for Water (Antoine Equation)**

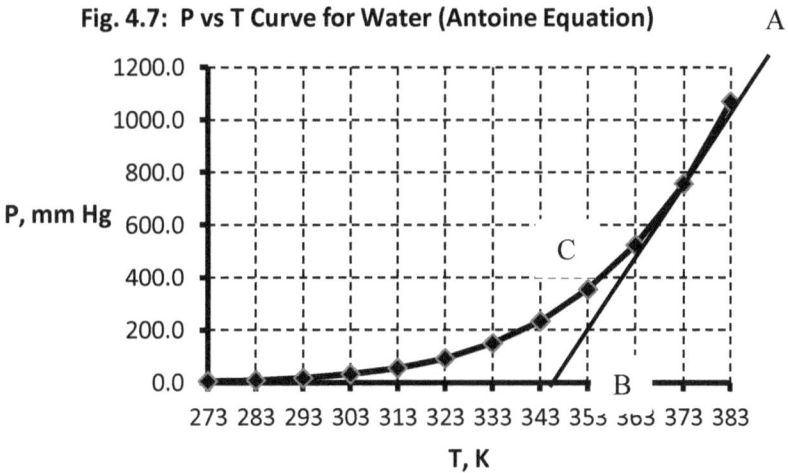

**Illustrative Example 4.24**

The Antoine equation for the prediction of the vapour pressures of pure substances is given by equation (4.45) above. If A = 18.3036, B = 3816.44 and C = - 46. 13 for water, determine its latent heat of vaporisation at 50 C and 1 atm using the Clausius-Clayperon and the Haggenmacher equations. The critical temperature and pressure for water are 647 K and 220.5 bar, respectively, and R = 8.314 J/mol K.

**Answer**

T = 50 C + 273 C = 323 K;   1 atm = 14.5 bar

$$P_R = \frac{14.5}{220.5} = 0.07 \quad T_R = \frac{323}{647} = 0.50 \quad \Delta Z = \left(1 - \frac{P_R}{T_R}\right)^{0.5} = \left(1 - \frac{0.07}{0.5}\right) = 0.93$$

Substituting these and the other given values in the Clausius – Clayperon and Haggenmacher equations we get, for the Clausius –

157

Clayperon equation

$$\Delta H_V = \frac{RBT^2}{(T+C)^2} = \frac{8.314 \ x \ 3816.44 \ x \ 323^2}{(323-46.13)^2} \ , \frac{J}{mol.K} \ .K. \frac{K^2}{K^2} = 43,183.88 \ \frac{J}{mol} \ Ans$$

and for the Haggenmacher equation

$$\Delta H_V = \frac{8.32 \ B \ \Delta Z \ T^2}{(T+C)^2} = \frac{8.32 \ x \ 3816.44 \ x \ 0.93 \ x \ 323^2}{(323-46.13)^2} \ , \frac{J}{mol.K} \ .K. \frac{K^2}{K^2}$$

$$= 40,190 \frac{J}{mol} \ Ans$$

The experimental value is 42,891 J/mol. The results above show that, for this situation, the Clausius –Clayperon equation gives a value that is +0.7 % in error, while the Haggenmacher equation gives a value that is -6.3 % in error, of the experimentally determined value.

### d) Use of Watson's Rule

This is a modification of Trouton's rule such that, if the latent heat of the substance is known at one temperature, say $\Delta H_{V_1}$ at $T_1$, it can be found at any other temperature, say $\Delta H_{V_2}$ at $T_2$. The Watson's equation is

$$\frac{\Delta H_{V_2}}{\Delta H_{V_1}} = \left( \frac{1-T_{R_2}}{1-T_{R_1}} \right)^{0.38} \qquad (4.100)$$

where $T_{R_1}$, and $T_{R_2}$ are the reduced temperatures under conditions 1 and 2.

### Illustrative Example 4.25

Suppose it is known that the latent heat of vaporisation of water, at 1 atm, 50 C, is 2382.84 kJ/kg. Determine its latent heat at 100 C, also, at 1 atm.

### Answer

For water, $T_C = 647.3$ K; $T_1 = 50 + 273 = 323$ K; $T_2 = 100 + 273 = 373$ K:

$$T_{R_1} = \frac{323}{647.3} = 0.50 ; \quad T_{R_2} = \frac{373}{647.3} = 0.58$$

From Watson's rule

$$\frac{\Delta H_{V_2}}{2382.84} = \left(\frac{1-T_{R_2}}{1-T_{R_1}}\right)^{0.38} = \left(\frac{1-0.58}{1-0.50}\right)^{0.38} = 0.94$$

Hence

$$\Delta H_{V_2} = 0.94 \times 2382.84 \ kJ/kg = 2239.9 \ kJ/kg \quad Ans.$$

The experimental value is 2260.2 kJ/kg. The predicted value is, therefore, -0.9 % in error.

e)      Use of Reference Substance Plots

Material property estimation by graphical procedures have gone out of fashion in recent times mainly because, compared to modern experimentally determined or analytical computer calculated values, they are tedious and not as accurate. For historical reasons, however, it is useful to report at least one procedure that was quite popular during its time. This is the procedure of reference substance plots.

Raoult's law is not of great use for solutions containing non-volatile solutes, because of the lowering of the vapour pressure of the volatile components and the elevation of boiling point of the solution caused by the presence of the non-volatile solute.

Recall the version of the Clausius-Clayperon equation (4.60) shown below

$$\frac{1}{P}\frac{dP}{dT} = \frac{\Delta H_V}{RT^2} \quad or \quad d\ln P = -\frac{\Delta H_V}{R}d\left(\frac{1}{T}\right) \qquad from \ (4.93)$$

If we choose water as a reference substance at a reference temperature, $T_W$, this equation becomes

$$d\ln P_W = -\frac{\Delta H_{V,W}}{R}d\left(\frac{1}{T_W}\right) \qquad (4.101)$$

For any other vaporisable liquid substance whose $\Delta H_V$ we would like to determine at the temperature T, this equation is

$$d\ln P = -\frac{\Delta H_V}{R}d\left(\frac{1}{T}\right) \qquad from \ (4.93)$$

The reference substance plot procedure consists of constructing charts in which either $1/T_W$ is plotted against $1/T$ or $\ln P$ versus $\ln P_W$ with either plot having a slope or gradient equal to $\Delta H_V/\Delta H_{VW}$. Since all the $T_W$, $P_W$ and $\Delta H_{VW}$ are known, $\Delta H_V$ can be calculated.

When P = Pw, the chart is known as the Duehring chart. From equations (4.93) and (4.101)

$$d \ln P_W = -\frac{\Delta H_{V,W}}{R} d\left(\frac{1}{T_W}\right) = d \ln P = -\frac{\Delta H_V}{R} d\left(\frac{1}{T}\right)$$

which, on integration, gives

$$\frac{1}{T_W} = \frac{\Delta H_V}{\Delta H_{V,W}} \frac{1}{T} + cons\tan t \qquad (4.102)$$

If the system obeys Trouton's rule, the constant is zero.

When T = Tw, the chart is known as the Cox chart. From equations (4.93) and (4.101)

$$\frac{R}{\Delta H_{V,W}} d \ln P_W = -d\left(\frac{1}{T_W}\right) = \frac{R}{\Delta H_V} d \ln P = -d\left(\frac{1}{T}\right)$$

which, on rearrangement and integration, gives

$$\ln P = \frac{\Delta H_V}{\Delta H_{V,W}} \ln P_W + cons\tan t \qquad (4.103)$$

The Duehring chart enables the experimental determination of the elevation af boiling point while the Cox chart helps to determine the lowering of the vapour pressure. The applicable equation is $P = k\ P_S$ where $P_S$ is the vapour pressure of pure solvent and k is a factor dependent on concentration. For an ideal solution, k = mole fraction of Raoult's law. The Duehring plot is an equal vapour pressure reference substance plot while the Cox plot is an equal -temperature reference substance plot.

To plot a Duehring point, the temperature at which pure water (the reference substance) has the same vapour pressure as that measured for the solution is plotted on the ordinate while the temperature of the solution is plotted on the abscissa.

For the Cox plot, the vapour pressure of pure water, obtained at the temperature of the solution, is plotted on the abscissa while the measured vapour pressure of the solution is plotted on the ordinate axis. They are, both, useful for determining latent heats of vaporisation for such mixtures.

**Illustrative Example 4.26**

Determine the latent heat of vaporization of an aqueous Ngu solution at

160

80 C using the Duerhing and Cox reference substance plots. The temperature-vapour pressure data were obtained as follows.

| Solution Temp, t, C | Solution Temp, T, K | Vapour Press of Soln, P, mm Hg | Satn Vap Press of Water at Soln Temp, $P_W$, mm Hg | Satn Temp of Water at Soln Vap Press, $T_W$, K |
|---|---|---|---|---|
| 60 | 333 | 105 | 147 | 326 |
| 70 | 343 | 173 | 231 | 336 |
| 80 | 353 | 305 | 351 | 349 |
| 90 | 363 | 477 | 523 | 361 |
| 100 | 373 | 705 | 755 | 371 |

Since the Duerhing plot is a plot of $1/T_W$ versus $1/T$, we can construct Table 1 from the data given as shown below.

**Table 1: Computed Data for the Duerhing Plot**

| Solution Temp, t, C | Solution Temp, T, K | Satn Temp of Water at Soln Vap Press, $T_W$, K | $1/T_W$ | $1/T$ |
|---|---|---|---|---|
| 60 | 333 | 326 | 0.00307 | 0.0030 |
| 70 | 343 | 336 | 0.00298 | 0.00292 |
| 80 | 353 | 349 | 0.00287 | 0.00283 |
| 90 | 363 | 361 | 0.00277 | 0.00276 |
| 100 | 373 | 371 | 0.00270 | 0.00268 |

This is plotted as shown in Fig 1 below.

The average slope is given by

$$\frac{\Delta H_V}{\Delta H_{VW}} = \frac{0.00307 - 0.00270}{0.0030 - 0.00268} = 1.15625 \tag{1}$$

Since $\Delta H_{VW}$ at 80 C is 2309 kJ/kg

$$\Delta H_V = 1.15625 \times 2309 = 2669.8 \, kJ / kg \tag{2}$$

For the Cox plot, we can construct a Table of ln P versus ln $P_W$ from the data given as shown below in Table 2. This is plotted as shown in Fig. 2 below.

The average slope is given by

161

$$\frac{\Delta H_V}{\Delta H_{VW}} = \frac{6.558 - 4.654}{6.627 - 4.990} = 1.1631 \qquad (3)$$

Since $\Delta H_{VW}$ at 80 C is 2309 kJ/kg

$$\Delta H_V = 1.1631 \times 2309 = 2685.6 \, kJ / kg \qquad (4)$$

The difference between the values predicted by the Cox and Duerhing plots is only 0.5 % and much less than the error of experimental errors

### Table 2:  Computed Data for the Cox Plot

| Solution Temp,  t,  C | Solution Temp, T, K | Ln P | Ln $P_W$ |
|---|---|---|---|
| 60 | 333 | 4.654 | 4.990 |
| 70 | 343 | 5.153 | 5.442 |
| 80 | 353 | 5.720 | 5.861 |
| 90 | 363 | 6.168 | 6.260 |
| 100 | 373 | 6.558 | 6.627 |

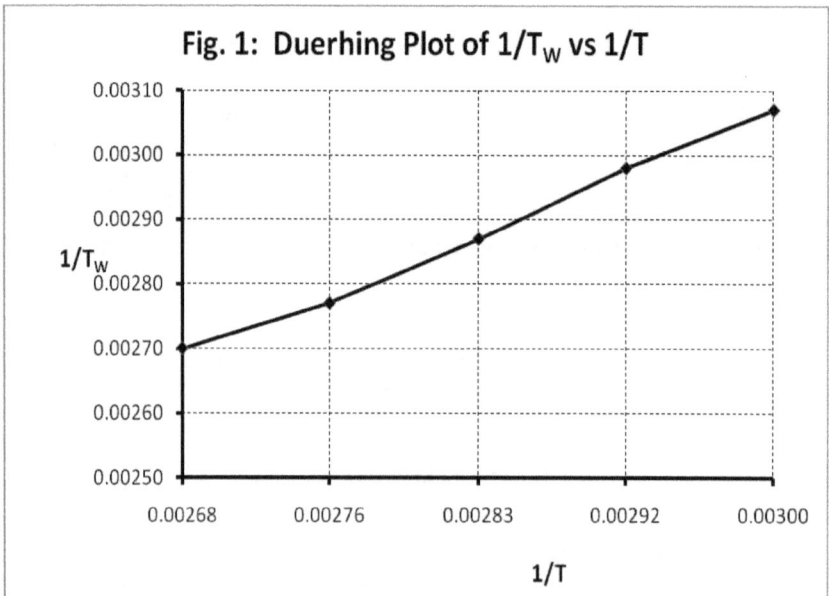

Fig. 1: Duerhing Plot of $1/T_W$ vs $1/T$

**Fig. 2: Cox Plot of ln P vs ln $P_w$**

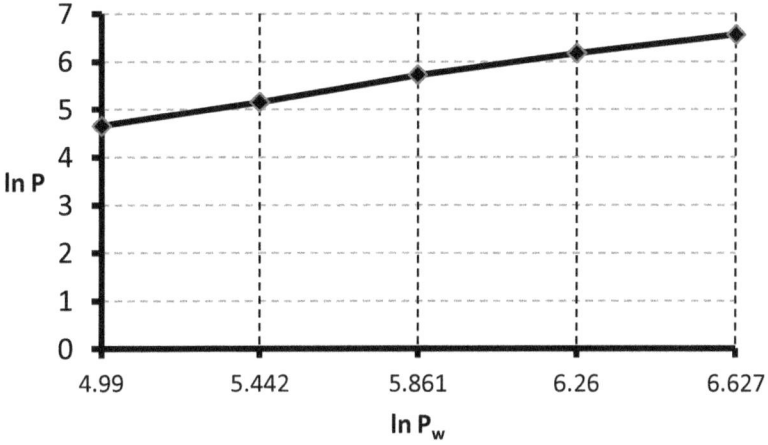

### 4.15.4: Prediction of Heats of Transition of Mixtures of Substances

If several pure liquid components make up a mixture, the latent heat of the mixture, $\Delta H_M$, is given by

$$\Delta H_V = x_1 \, \Delta H_{V1} + x_2 \, \Delta H_{V2} + x_3 \, \Delta H_{V3} + \ldots\ldots \qquad (4.104)$$

where $x_1$, $x_2$, $x_3$, etc are the mole fractions of the components in the mixture and $\Delta H_{V1}, \Delta H_{V2}, \Delta H_{V3}$, etc., are their respective latent heats at the given temperature.

The boiling point of the mixture is the molal average boiling point (MABP) at that pressure given by the condition when

$$\sum_i K_i \, x_i = 1.0 \qquad (4.105)$$

where

$$K_i = \frac{\gamma_i \, P_i^o}{P_T} \qquad (4.106)$$

$K_i$ is the equilibrium constant; $\gamma_i$ the activity coefficient of component, $i$; $P_i^o$ the vapour pressure of component, $i$, and $P_T$ the system total pressure. $P_i^o$ is estimated from the Antoine equation while $\gamma_i$ is estimated from the Wilson equation (Coulson, Richardson & Sinnott, 1983) as

163

$$\ln \gamma_i = 1.0 - \ln\left(\sum_{j=1}^{N} x_j A_{kj}\right) - \sum_{j=1}^{N}\left(\frac{x_i A_{ik}}{\sum_{j=1}^{N} x_j A_{ij}}\right) \qquad (4.107)$$

where $\gamma_k$ is the activity coefficient of component, k; $A_{ij}$ and $A_{ik}$ the Wilson coefficients for the binary pair, $i$ and $j$; and N is the number of components.

**Illustrative Example 4.27**

Determine the latent heat of vaporisation of a mixture of ethanol and water containing 45 mole % ethanol. What will be the boiling point of this mixture at 1 atmosphere?

**Answer**

Since the latent heat of the mixture cannot be determined unless its boiling point is known, the latter has to be estimated first, using equation the Antoine equation. The pure component vapour pressures are determined as a function of temperature, using the Antoine equation. The Antoine equation constants are

| Component | Identification Subscript | A | B | C |
|---|---|---|---|---|
| Ethanol | 1 | 18.9119 | 3803.98 | - 41.68 |
| Water | 2 | 18.3036 | 3816.44 | - 46.13 |

Hence, for ethanol, the pure component vapour pressure is given by

$$\ln p_1^o = A - \frac{B}{T+C} = 18.9119 - \frac{3803.98}{T - 41.68}$$

or

$$p_1^o = \exp\left(18.9119 - \frac{3803.98}{T - 41.68}\right) \qquad (1)$$

while for water, the pure component vapour pressure is

$$\ln p = A - \frac{B}{T+C} = 18.3036 - \frac{3816.44}{T - 46.13}$$

or

164

$$p_2^o = \exp\left(18.3036 - \frac{3816.44}{T - 46.13}\right) \qquad (2)$$

$\gamma_k$ is estimated from the Wilson equation, (4.94), as follows. The $A_{ij}$ and $A_{ji}$ values for ethanol, $i$, and water, $j$, are given by Hirata (Coulson, Richardson & Sinnott, (1983) as

| Component | Ethanol | Water |
|-----------|---------|-------|
| Ethanol | 1 | 0.1108 |
| Water | 0.9560 | 1 |

For ethanol, $k = 1$ and $N = 2$. From equations (4.102)

$$\ln \gamma_1 = 1.0 - \ln\left(x_1 A_{11} + x_2 A_{12}\right) - \left(\frac{x_1 A_{11}}{x_1 A_{11} + x_2 A_{12}} + \frac{x_2 A_{21}}{x_1 A_{21} + x_2 A_{22}}\right)$$

$$= 1.0 - \ln(0.45 \, x \, 1 + 0.55 \, x \, 0.1108)$$

$$-\left(\frac{0.45 \, x \, 1}{0.45 \, x \, 1 + 0.55 \, x \, 0.1108} + \frac{0.55 \, x \, 0.9560}{0.45 \, x \, 0.9560 + 0.55 \, x \, 1}\right)$$

$$= 1.0 + 0.6715 - (0.8807 + 0.5364) = 0.2544$$

From which we get that $\gamma_1 = 1.2897$.

Similarly, for water, $k = 2$ and $N = 2$. From equations (4.94)

$$\ln \gamma_2 = 1.0 - \ln\left(x_1 A_{21} + x_2 A_{22}\right) - \left(\frac{x_1 A_{12}}{x_1 A_{11} + x_2 A_{12}} + \frac{x_2 A_{22}}{x_1 A_{21} + x_2 A_{22}}\right)$$

$$= 1.0 - \ln(0.45 \, x \, 10.9560 + 0.55 \, x \, 1)$$

$$-\left(\frac{0.45 \, x \, 0.1108}{0.45 \, x \, 1 + 0.55 \, x \, 0.1108} + \frac{0.55 \, x \, 1}{0.45 \, x \, 0.9560 + 0.55 \, x \, 1}\right)$$

$$= 1.0 + 0.0120 - (0.0976 + 0.5611) = 0.3533$$

From which we get that $\gamma_2 = 1.4238$.

When we substitute the values of $\gamma_1$, $P_1^o$ and $P_T$ in equations (4.101) we get

$$K_1 = \frac{\gamma_1 P_1^o}{P_T} = \frac{1.2897 \exp\left(18.9119 - \dfrac{3803.98}{T - 41.68}\right)}{760} = 277337.7 \exp\left(-\frac{3803.98}{T - 41.68}\right) \qquad (3)$$

Similarly with the values of $\gamma_2$, $P_2^o$ and $P_T$ in equations (4.101) we get

$$K_2 = \frac{\gamma_2 P_2^o}{P_T} = \frac{1.4238 \exp\left(18.3036 - \dfrac{3816.44}{T - 46.13}\right)}{760} = 166643.3 \exp\left(-\frac{3816.44}{T - 46.13}\right) \qquad (4)$$

Substituting equations (3) and (4) in equation (4.100)

$$\sum_i K_i x_i = 1.0 = 0.45 \times 277337.7 \exp\left(-\frac{3803.98}{T-41.68}\right)$$

$$+ 0.55 \times 166643.3 \exp\left(-\frac{3816.44}{T-46.13}\right)$$

$$= 124802 \exp\left(-\frac{3803.98}{T-41.68}\right) + 91653.8 \exp\left(-\frac{3816.44}{T-46.13}\right) \quad (5)$$

The boiling point is found by trial and error. We, first, assume a boiling point and use it in equation (5). If it is not equal to 1, we assume another value until we get the sum equal to 1.

For example, if t = 80 C or T = 273 +80 = 353 K, equation (5) becomes

$$\sum_i K_i x_i = 124802 \exp\left(-\frac{3803.98}{353-41.68}\right) + 91653.8 \exp\left(-\frac{3816.44}{353-46.13}\right)$$

$$= 124802 \times 4.94 \times 10^{-6} + 91653.8 \times 3.97 \times 10^{-6} = 0.98 \quad (6)$$

If t = 80.5 C or T = 273 +80.5 = 353.5 K, equation (5) becomes

$$\sum_i K_i x_i = 124802 \exp\left(-\frac{3803.98}{353.5-41.68}\right) + 91653.8 \exp\left(-\frac{3816.44}{353.5-46.13}\right)$$

$$= 124802 \times 5.03 \times 10^{-6} + 91653.8 \times 4.05 \times 10^{-6} = 0.999 \approx 1.0 \quad (7)$$

The latent heats of vaporisation of the components, at their normal boiling points, are obtained from Tables as follows

| Component | Boiling Point, C | $\Delta H_V$, kJ/kmol |
|---|---|---|
| Ethanol | 78.3 | 38,770 |
| Water | 100 | 40,683 |

These values are used in the Watson's equation (4.100) to estimate the latent heat of each component at the temperature of the mixture, 80.5 C. Thus for water at 80.5 C, with $\Delta H_V$ = 40,683 kJ/kmol and $T_C$ = 647.3 K, the reduced temperatures $T_{R1}$ at 373K is 373/647.3 = 0.58 and $T_{R2}$ at 353.5K is 0.55. Hence

$$\Delta H_{V_2} = \Delta H_{V_1}\left(\frac{1-T_{R_2}}{1-T_{R_1}}\right)^{0.38} = 40,683\left(\frac{1-0.55}{1-0.58}\right)^{0.38} = 41,763.7 \, kJ/kmol \quad (8)$$

Similarly, for ethanol at 80.5 C, with $\Delta H_V$ = 38,770 kJ/kmol and $T_C$ = 516.2 K, the reduced temperatures $T_{R1}$ at 351.3K is 351.3/516.2 = 0.68

166

and $T_{R2}$ at 353.5K is 0.69.

$$\Delta H_{V_2} = \Delta H_{V_1} \left(\frac{1-T_{R_2}}{1-T_{R_1}}\right)^{0.38} = 38,770 \left(\frac{1-0.68}{1-0.69}\right)^{0.38} = 39,240.6 \, kJ \, / \, kmol \quad (9)$$

Hence, from equation (4.100), the latent heat of vaporisation of the mixture is $\Delta H_V = x_1 \Delta H_{V1} + x_2 \Delta H_{V2}$

$$= 0.45 \, x \, 39,240.6 + 0.55 \, x \, 41,763.7 = 40,178.3 \, kJ/mol \quad Ans.$$

## 4.16: Thermodynamic Descriptions of Chemical Equilibrium

While phase equilibrium studies enable us to determine relationships between physical variables in non-reacting systems, chemical equilibrium studies enable us to do the same for systems in which chemical reactions occur. Real life situations are such, also, that we, often, have to contend, simultaneously, with both phase and chemical equilibrium in the same system.

We can study chemical equilibrium separately, by keeping conditions of phase equilibrium constant as we did phase equilibrium by assuming no chemical reactions. For systems in which chemical reactions occur, two parameters, the equilibrium constant and the heat of reaction, provide very effective descriptuions of chemical equilibrium in that system.

### 4.16.1: IUPAC Convention for Chemical Reaction Studies

The IUPAC convention is a more complicated protocol than can be presented here. The essential point, however, is that it specifies that, for a chemical reaction in which $v_A$ moles of component A reacts with $v_B$ moles of component B to produce $v_C$ moles of component C and $v_D$ moles of component D such that the equation of reaction is

$$v_A A + v_B B \leftrightarrow v_C C + v_D D \qquad (4.108)$$

1. A and B are the reactants
2. C and D are the products
3. any change as a result of the reaction will be that between the final condition minus the intial condition, where the final condition is written on the right hand side of the equation and the intial condition on the left hand side of the equation.
4. $v_A$, $v_B$, $v_C$, $v_D$ are the stoichiometric coefficients such that $v_A$, $v_B$, are negative and $v_C$, $v_D$ are positive

## Reactant and Product Quantification

The formula of any substance appearing in a reaction equation indicates two things, the type of substance and its quantity in mols (moles) or kmols (kmoles). For example

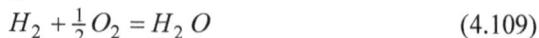

$$H_2 + \tfrac{1}{2}O_2 = H_2 O \qquad (4.109)$$

indicates that 1 mol of hydrogen reacts with half a mol of oxygen to produce 1 mol of water.

## The Physical State of Reactants and Products

It is, always, assumed.that the substances staking part in a reaction as shown in the chemical reaction equation, atre in their normal states of occurrence in nature. Where their state is diffrenet from this or confusion may arise, symbols are used, next to their chemical formulae, to indicatwe as follows; (g) for the gaseous state, (l) for the liquid state and (s) for solid state.

For example, $H_2O$ (g) means 1 mole of water vapour, $H_2O$ (l) means 1 mole of liquid water while $H_2O$ (s) means 1 mol of ice. For polymorphous substances, further clarification is necessary such as say for carbon as C (diamond), C (graphite) or for sulphur as S (rhombic), ice as ice (II), etc.

## Concentration of Reactants and Products

In addition to the usual definitions of concentration in the liquid and gas phases, certain notations have been developed by chemists to enable cryptic presentations of concentration. These are, taking the concentration of 1 mole of HCl, for example.

| How Concentration is described | What it means |
| --- | --- |
| HCl (m = 1) | molarity of HCl is 1; that is 1 gmol of HCl in 1000 cc of solution |
| HCl (o = 0.97) | molality of HCl is 0.97; that is 0.97 gmol of HCL in 1000 cc of solvent |
| HCl (x = 0.02) | mole fraction of HCl is 0.02; that is 2 gmols of HCl in 100 solution or 20 gmols of HCl in 1000 gmols of solution |
| HCl (n = 56) | there are 56 gmols of solvent per gmol of HCl |

When the solution is so dilute that additional dilution produces no thermal effect; the concentration is written, for HCl as an example, as HCl (aq). For non-aqeous solvents, the formula for the solvent is included.

For example, when 1 gmol of HCl is dissolved in 56 gmols of chloroform, the concentration is written as HCl ($CHCl_3$, n = 56) or HCl ($CHCl_3$, x = 0.5), etc.

### 4.16.2: Thermal Changes associated with Chemical Reactions

Every chemical reaction is associated with one or more of the following thermal changes which are named according to their character or phenomenon driving the reaction. For standardization purposes, each thermal change is defined for the reaction occuring at 1 atm pressure and in which the reactants and products are, unless otherwise specified, in their normal states of existence.

Standard Heat of Reaction

The standard heat of reaction is defined as the change in enthaJpy when reaction occurs as written at 1 atm., with reactants and products at the constant temperature of 25 C. For example, the standard heat of reaction, $\Delta H_R^{298}$, for the dissolution of zinc in dilute HCl is given as

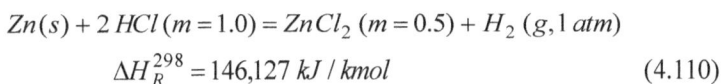

$$Zn(s) + 2\,HCl\,(m = 1.0) = ZnCl_2\,(m = 0.5) + H_2\,(g, 1\ atm)$$
$$\Delta H_R^{298} = 146{,}127\ kJ\,/\,kmol \qquad (4.110)$$

When the system absorbs energy as a result of the reaction, the reaction is said to be endothermic and the sign of the heat of reaction is positive. When the system loses energy as a result of the reaction, the reaction is said to be exothermic and the sign of the heat of reaction is negative.

The heat of reaction depends, also, on the chemical and physical nature of the reactants and products and on the conditions under which the reaction takes place.

Effect of Temperature on the Heat of Reaction

Most reactions of industrial or commercial importance do not take place at 1 atm and 25 C. It is necessary, therefore, to be able to convert the standard heat of reaction to the actual heat of reaction at the actual reaction temperature.

Suppose we wished to estimate the heat of reaction, $\Delta H_R^{T_2}$, at the temperature, $T_2$, for a reaction which has a heat of reaction at $T_1$ of $\Delta H_R^{T_1}$. Since the definition of the heat of reaction requires that all reactants and products enter and leave at $T_2$, for the calculated $\Delta H_R^{T_2}$ to be the correct one, both the reactants and products must be heated from $T_1$ to $T_2$.

The heat of reaction at $T_2, \Delta H_R^{T_2}$ must then be equal to the heat of reaction at $T_1$ of $\Delta H_R^{T_1}$ plus the net change in energy required to bring the reactants and products from $T_1$ to $T_2$ (Hess's Law). That is

$$\Delta H_R^{T_2} = \Delta H_R^{T_1} + \sum_{j \, products} Cp_j \left(T_2 - T_1\right) - \sum_{i \, reactants} Cp_i \left(T_2 - T_1\right) \qquad (4.111)$$

This can be rearranged as

$$\frac{\Delta H_R^{T_2} - \Delta H_R^{T_1}}{\left(T_2 - T_1\right)} = \sum_{j \, products} Cp_j - \sum_{i \, reactants} Cp_i = \Delta Cp \qquad (4.112)$$

In the limit as $(T_2 - T_1)$ tends to zero

$$\frac{d \Delta H_R^T}{d T} = \Delta Cp \qquad (4.113)$$

Equation (4.113) is known as the Kirchoff's equation. It can be integrated into its more common form

$$\Delta H_R^{T_2} = \Delta H_R^{T_1} + \int_{T_1}^{T_2} \Delta Cp \, d T \qquad (4.114)$$

For most gases, for which

$$Cp = a + bT + cT^2 + \dots \qquad (4.115)$$

where a, b, c, etc are constants,

$$\Delta H_R^{T_2} = \Delta H_R^{T_1} + a\left(T_2 - T_1\right) + \frac{b \left(T_2^2 - T_1^2\right)}{2} + \frac{c \left(T_2^3 - T_1^3\right)}{3} + \dots \qquad (4.116)$$

## Heat of Formation

This is the enthalpy change associated with the formation of a compound or substance from its elements at 25 C and 1 atm pressure, the elements being in their normal state of occurrence under those conditions.

For reasons of standardisation, the standard heat of formation is defined with respect to one mol or kmol of substance or compound and is variously written as $\Delta H_f^{298}$, $\Delta H_{298}^f$ or as $\Delta H_{298}^o$.

When the heat of formation is negative, the substance is regarded as an exothermc substance while if it is positive, the substance is an endothermic substance. Elements in their natrural state of occurrence at 25 C, 1 atm, such as $H_2$, $O_2$, $N_2$, have zero heat of formation.

When an element exhibits allotropy, one allotrope is selected to have zero heat of formation while the heats of formation of the others are based on it.

Heats of formation form the fundamental bases for the calculation of the heats of reaction, combustion, etc.

Heat of Combustion

The standard heat of combustion is the heat of reaction resulting from the oxidation of a substance with molecular oxygen, with the reactants and products in their standard states at 25 C and 1 atm. The actual value of the standard heat of combustion depends on the extent to which oxidation is carried and on the final state of the products of combustion.

The extent to which combustion is carried is specified in the terms of whether there was complete combustion (known as stoichiometric combustion), or there was incomplete or partial combustion. This depends, in turn, on the oxygen to fuel ratio (air to fuel ratio, if air is the source of oxygen).

The final state of the products is determined by whether the $H_2O$, produced during combustion, is left in the product gases as water vapour or condensed out as liquid water. For hydrocarbons, the standard state is defined differently. Instead of 25 C, 1 atm, the standard state is 15.5 C (60 F) and 1 atm.

Two kinds of standard heats of combustion are in use, especially, in the analysis of the combustion of hydrocarbons. These are the gross or higher heat of combustion, also known as the Higher Heating Value

(HHV) and the net or lower heat of combustion, known as the Net or Lower Heating Value (NHV, LHV).

Both of these definitions apply to complete combustion with the HHV applying to the situation in which the water vapour, produced during combustion, is left in the product gases as water vapour while the LHV is applied to the situation in which the water vapour is condensed out of the product gases as liquid water.

## Heat of Neutralisation

This is the heat of reaction when neutralisation occurs between an acid and a base. The heat of neutralisation is, normally, calculated from the heats of formation of the various species involved and is found to be constant for strong acids and bases. Take, for example, the neutralization reaction between NaOH and HCl, a strong base and a strong acid. The reaction may be written as

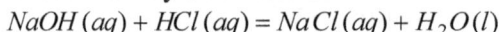

$$NaOH\,(aq) + HCl\,(aq) = NaCl\,(aq) + H_2O\,(l)$$

The actual reaction, according to Arrhenius, is really that between $H^+$ and $OH^-$ ions as

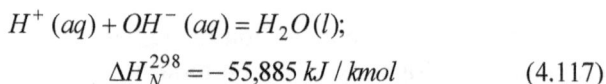

$$H^+\,(aq) + OH^-\,(aq) = H_2O\,(l);$$

$$\Delta H_N^{298} = -55{,}885\ kJ\,/\,kmol \tag{4.117}$$

This suggests that, for strong acids and bases, the heat of neutralization should be independent of the identities of the acid or base.

The heat of neutralisation is found to be zero, however, when dilute, aqeous solutions of neutral salts of strong acids and bases are mixed.

For example, the reaction

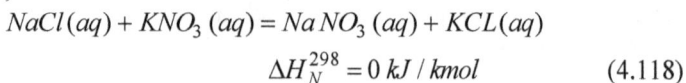

$$NaCl\,(aq) + KNO_3\,(aq) = NaNO_3\,(aq) + KCL\,(aq)$$

$$\Delta H_N^{298} = 0\ kJ\,/\,kmol \tag{4.118}$$

illustrates the so called principle of thermoneutrality of solutions.

## Heat of Solution

This is the heat absorbed or evolved when one mol of material dissolves in another to form a solution. It will include the heats of solvation, ionisation and hydration if any or all of these are involved in the solution forming process. It is, generally, presented either as the integral heat of solution or as the differential heat of solution.

In either case, the heats of solution of neutral, non-dissociating salts are, generally, positive while those of ionisable salts are negative. The heats of solution of organic and inorganic compounds in water can be large, especially, for the strong mineral acids and alkalies.

The numerical value of the heat of solution depends on the amount and nature of the solvent and solute, as well as on the initial and final temperature and concentration of the solution (Hougen et al., 1954). When both the materials forming the solution are liquids, the heat of solution is known as the heat of mixing. For organic solutions, the heat of mixing is, usually, small compared to other thermal quantities.

Integral Heat of Solution

The standard, integral heat of solution with respect to component, i, at a particular concentration, is defined as the change in enthalpy when one mol of solute, i, is dissolved in enough pure solvent to produce a solution of the given concentration, all materials entering and leaving at 298 K and one atm. Its value depends on the number of moles of i and will increase as i increases to a maximum asymptotic value known as the integral heat of solution at infinite dilution.

In practice, infinite dilution can be assumed for many solutions if the solvent to solute mole ratios fall in the range of 20 to 200. In those situations, for example, the integral heat of solution for the dissolution of a component in a solvent will be equal to the enthalpy change for the process (Daniels et al, 1962).

When a liquid solute dissolves in a solvent to form an ideal solution, the integral heat of solution is zero. When a solid solute dissolves in a solvent to form an ideal solution, the integral heat of solution is equal to the molar heat of fusion required to produce a supercooled liquid at the temperature of the solution (Daniels et al, 1962). Figs 4.8 and 4.9 show charts of integral heats of solution of some acids and alkalis in water.

The integral heat of solution between two molalities, $m_1$ and $m_2$, is given by

$$\Delta H_{IHS}^{298} = \Delta H_{HS2}^{298} - \Delta H_{HS1}^{298}, \; kJ / kmol \qquad (4.119)$$

**Fig. 4.8: Integral Heats of Solution of Some Acids in Water
(Hougen et al, 1954)**

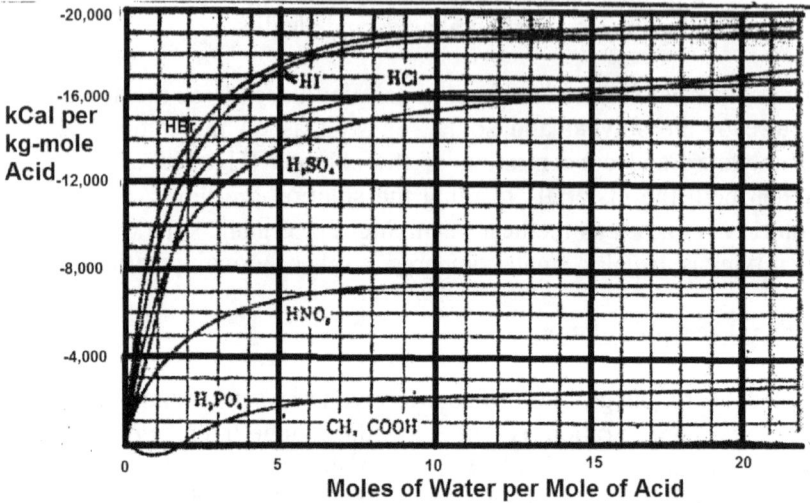

Moles of Water per Mole of Acid

## Differential Heat of Solution

The differential heat of solution is the incremental change in the
enthalpy of a solution as a result of the addition of one mol of solute, i,
to so large a quantity of solution that the addition of that one more mole
of solute does not, appreciably, change the molality of the solution.
The differential heat of solution is

$$\Delta H_{DHS}^{298} = \left( \frac{\partial \Delta H_{HS}}{\partial n_i} \right)_{P,T,n_j} \tag{4.120}$$

This can be reduced to measurable variables, m and $\Delta H_{IHS}$. Since the
addition of dm moles of solute, at constant T and P, to a solution
already containing m moles of solute in 1000 g of solvent (molality, m)
produces the enthalpy change of $d(m\Delta H_{IHS})$ the differential heat of
solution for solute, i, is

$$\Delta H_{DHS}^{298} = \left( \frac{\partial \Delta H_{HS}}{\partial n_i} \right)_{P,T,n_j} = \left( \frac{\partial m \Delta H_{IHS}}{\partial m} \right)_{P,T}$$

$$= \Delta H_{IHS} + m \left( \frac{\partial \Delta H_{IHS}}{\partial m} \right)_{P,T} \tag{4.121}$$

174

**Fig. 4.9: Integral Heats of Solution of Some Alkalis in Water**
**(Hougen et al, 1954)**

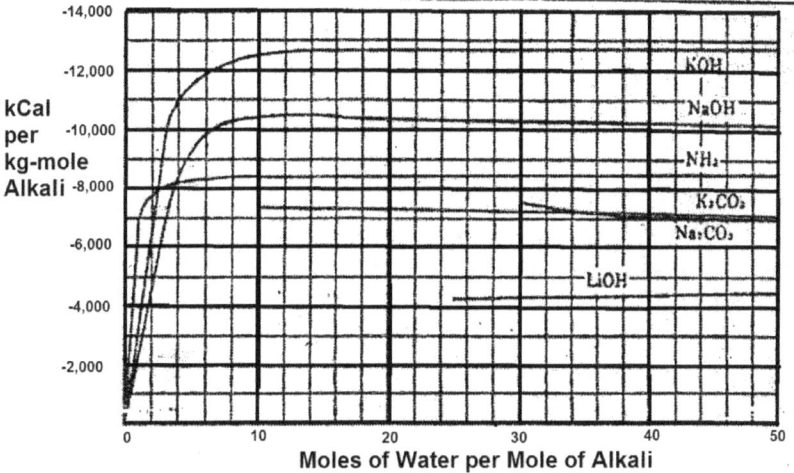

## Heats of Ionisation and Dissociation

These are the heat energy changes associated with the ionisation or dissociation of an element or compound.

For an element

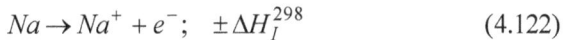

$$Na \rightarrow Na^+ + e^-; \quad \pm \Delta H_I^{298} \tag{4.122}$$

For a compound

$$Na\,OH \rightarrow Na^+ + OH^-; \quad \pm \Delta H_D^{298} \tag{4.123}$$

## Heats of Solvation and Hydration

Certain solutes can form definite chemical compounds with their solvents in a process known as solvation. The enthalpy change associated with this process is the heat of solvation. When the solvent is water, the heat of solvation is referred to as the heat. of hydration.

## Enthalpy Concentration Charts

Enthalpy concentration charts are very useful in determining the enthalpy of solutions formed by mixing other solutions. Its use is based on the modified form of equation (4.104) in order to account for the integral heat of solution of the solutes in the solutions being mixed. For a binary system, this becomes

175

$$H_S = x_1 H_1 + x_2 H_2 + x_2 \Delta H_{IHS2} \qquad (4.124)$$

where $x_1$ and $x_2$ are not mass fractions in the normal sense but the mass of solute in solutions 1 and 2 respectively as fractions of their combined mass in the final solution. $H_1$ and $H_2$ are the enthalpies of the pure component in the solutions and $\Delta H_{IHS2}$, the integral heat of solution of the solute.

When the solute is a liquid, it is normal to take the reference temperature as 0 C. When it is a solid, however, the reference temperature used is that set by the industry. In the pulp and paper industry, for example, the reference temperature suggested for black liquor solutions is 80 C (Stoy & Fricke, 1994).

Typical enthalpy- concentration charts, such as those for HCl and $H_2SO_4$, are shown in Fig 4.10.

### 4.16.3: The Equilibrium Constant

The equilibrium constant is based on the general thermodynamic premise that all natural processes always tend to a state of minimum energy, or for some purposes, a state of higher disorder. For chemical reactions, specifically, this means that at chemical equilibrium, the change in the Gibbs free energy must be zero so that the rate of the forward reaction must be equal to the rate of the backward reaction. That is

$$\Delta G = \sum_{j\ products} G_j - \sum_{i\ reactants} G_i = 0 \qquad (4.125)$$

Since the rate of chemical reaction is expressed in the form

$$Rate = Rate\ cons\tan t\ x\ reaction\ driving\ force \qquad (4.126)$$

where the reaction driving force may be expressed as concentrations, pressures, mole or fractions of the chemical constituents.

Thus when the forward and backward rates of reaction are equal, it can be seen that

$$\frac{forward\ reaction\ rate\ cons\tan t}{backard\ reaction\ rate\ cons\tan t} = \frac{driving\ force\ for\ products}{driving\ force\ for\ reac\tan ts} \qquad (4.127)$$

The equilibrium constant is defined as the ratio of the forward to the backward rate constants. Because this ratio must be expressed in measurable variables, we can either use the kinetics in the case of a particular reaction or chemical thermodynamics for more general relationships for the equilibrium constant.

From kinetics, for the ideal gases under consideration, the rate of the forward reaction, $R_F$, is

$$R_F = k_F \, P_A^{V_A} . P_B^{V_B} \tag{4.128}$$

For the backward reaction, $R_B$, the rate is

$$R_B = -k_B \, P_C^{V_C} . P_D^{V_D} \tag{4.129}$$

At chemical equilibrium, $R_F = - R_B$. That is

$$k_F \, P_A^{V_A} . P_B^{V_B} = k_B \, P_C^{V_C} . P_D^{V_D} \tag{4.130}$$

### Fig. 4.10: Enthalpy-Concentration Charts for HCL and $H_2SO_4$ (Hougen et al, (1954)

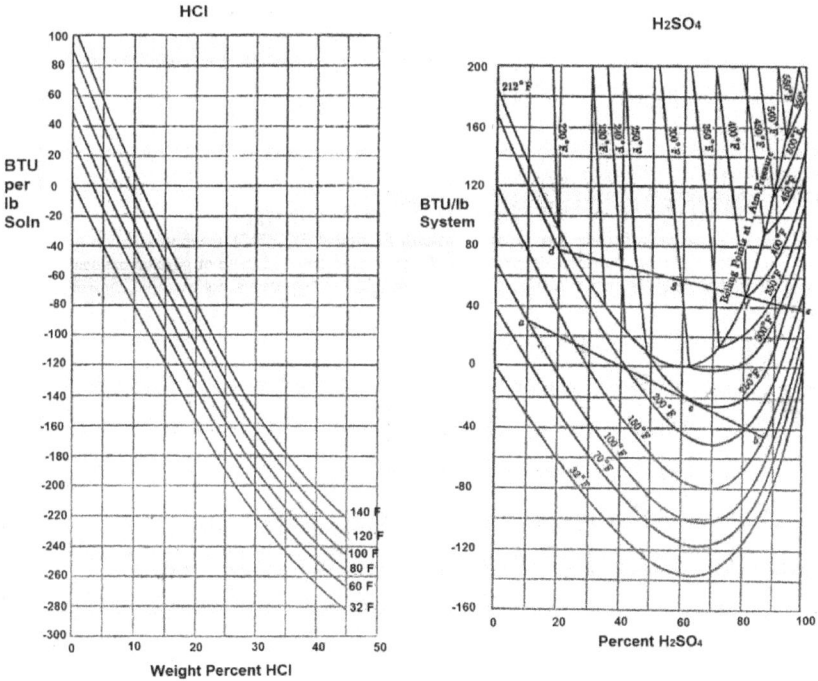

The equilibrium constant, $K_P$, based on pressure, is, therefore,

$$K_P = \frac{k_F}{k_B} = \frac{P_C^{V_C} . P_D^{V_D}}{P_A^{V_A} . P_B^{V_B}} \tag{4.131}$$

Can we evaluate $K_P$ without recourse to expensive experimentation? The Gibbs-Duhem equation is useful in this respect and, for any component in a closed system in phase equilibrium, is stated as

$$dG = -S \, dT + V \, dP \tag{4.132}$$

At constant temperature,
$$dG = V \, dP \tag{4.133}$$
For one mole of an ideal gas,
$$V = \frac{RT}{P} \tag{4.134}$$
so that, from (4.134) and (4.133)
$$dG = \frac{RT}{P} d P \tag{4.135}$$
Equation (4.127) can be integrated between an initial condition, 1, and a final condition, 2, to obtain

$$\Delta G = G_2 - G_1 = RT \ln\left(\frac{P_2}{P_1}\right) \tag{4.136}$$

To give this equation universal application, we may chose to define the initial condition with respect to some standard reference state.

For chemical reactions, this standard reference state has been chosen to be 1 atmosphere pressure and temperature of 25 C or 298 K with variables at this state designated with either an $o$ or 298 as superscript or subscript. We shall choose o as the superscript in this book.

Thus, between the standard state and any other state at a temperature, T, pressure, P, and Gibb's free energy, G, equation (4.136) becomes

$$G = G^o + RT \ln\left(\frac{P}{P^o}\right) = G^o + RT \ln P \tag{4.137}$$

since $\ln P^o = \ln 1.0 = 0$ if $P^o$ is in atm.

We can, now, apply equation (4.137) to each of the components in the chemical reaction depicted by equation (4.108). Thus from equation (4.137), for reactant A, with partial pressure, $P_A$, in the mixture,
$$G_A = G_A^o + RT \ln P_A \tag{4.138}$$
Similarly for reactant B, with partial pressure, $P_B$, in the mixture,
$$G_B = G_B^o + RT \ln P_B \tag{4.139}$$
and for product C, with partial pressure, $P_C$, in the mixture,
$$G_C = G_C^o + RT \ln P_C \tag{4.140}$$
and for product D, with partial pressure, $P_D$, in the mixture,
$$G_D = G_D^o + RT \ln P_D \tag{4.141}$$
From equations (4.109), (4.138) to (4.141) and (4.125) we can get that
$$\Delta G_R = v_C G_C + v_D G_D - v_A G_A - v_B G_B$$

178

$$= \Delta G_R^o + RT \ln \frac{P_C^{v_C} . P_D^{v_D}}{P_A^{v_A} . P_B^{v_B}} = \Delta G_R^o + RT \ln K_P \tag{4.142}$$

where

$$\Delta G_R^o = v_C G_C^o + v_D G_D^o - v_A G_A^o - v_B G_B^o \tag{4.143}$$

Thus, from (4.142) when $\Delta G_R = 0$

$$-\Delta G_R^o = RT \ln \frac{P_C^{v_C} . P_D^{v_D}}{P_A^{v_A} . P_B^{v_B}} = RT \ln K_P \tag{4.144}$$

To express equation (4.144) in terms of other forms of reaction driving force such as concentration or mole fractions, recall that for an ideal gas, concentration, C, can be expressed as

$$C = \frac{n}{V} = \frac{P}{RT} \tag{4.145}$$

while mole fraction, y, can be expressed as

$$y_i = \frac{P_i}{P_T} \tag{4.146}$$

where $n_i$ is the number of moles of component, i, V is the volume of the mixture and $P_i$ is the partial pressure of component, i. When these are substituted in equation (4.144) we get that

$$K_P = \frac{P_C^{v_C} . P_D^{v_D}}{P_A^{v_A} . P_B^{v_B}} = \frac{(C_C RT)^{v_C} . (C_D RT)^{v_D}}{(C_A RT)^{v_A} . (C_B RT)^{v_B}}$$

$$= \frac{C_C^{v_C} . C_D^{v_D}}{C_A^{v_A} . C_B^{v_B}} . (RT)^{(v_C + v_D - v_A - v_B)} = K_C . (RT)^{(v_C + v_D - v_A - v_B)} \tag{4.147}$$

where $K_C$ is the equilibrium constant expressed on the basis of molar concentration. Similarly

$$K_y = \frac{P_C^{v_C} . P_D^{v_D}}{P_A^{v_A} . P_B^{v_B}} = \frac{(y_C P_T)^{v_C} . (y_D P_T)^{v_D}}{(y_A P_T)^{v_A} . (y_B P_T)^{v_B}}$$

$$= \frac{y_C^{v_C} . y_D^{v_D}}{y_A^{v_A} . y_B^{v_B}} . (P_T)^{(v_C + v_D - v_A - v_B)} = K_y . (P_T)^{(v_C + v_D - v_A - v_B)} \tag{4.148}$$

where $K_y$ is the equilibrium constant expressed on the basis of mole fraction.

So far, the equilibrium constant has been evaluated at constant sytem temperature and constant total pressure. How is the equilibrium constant affected by a change in one or both of these in the system?

## Effect of Pressure on the Equilibrium Constant

Equation (4.147) shows that $K_P$ and $K_C$ would be independent of pressure because each is a ratio of component pressures or concentrations. $K_y$, however, as equation (4.148) illustrates, depends on the total pressure of the system and, because partial pressures are involved, would, also, depend on the presence of inert gases in the system.

## Effect of Temperature on the Equilibrium Constant

The effect of temperature is not as obvious as that of pressure but may be ascertained by differentiating equation (4.144) with respect to temperature. That is

$$\frac{\partial \Delta G_R^o}{\partial T} = -RT \frac{\partial \ln K_P}{\partial T} \tag{4.149}$$

But

$$d G_R = -S_R \, dT + V \, dP \tag{4.150}$$

which, on differentiation with the pressure constant, becomes

$$\frac{\partial G_R}{\partial T} = -S_R = \frac{G_R - H_R}{T} \tag{4.151}$$

since

$$G_R = H_R - T S_R \tag{4.152}$$

Equation (4.151) may also be stated as

$$\frac{\partial \Delta G_R}{\partial T} = -\Delta S_R = \frac{\Delta G_R - \Delta H_R}{T} \tag{4.153}$$

in which form it is known as the Gibbs-Helmholtz equation. At equilibrium, $\Delta G_R = 0$ and equation (4.153) becomes

$$\frac{\partial \Delta G_R}{\partial T} = -\Delta S_R = -\frac{\Delta H_R}{T} \tag{4.154}$$

Equation (4.154) is also true at the standard reference state so that we can say that

$$\frac{\partial \Delta G_R^o}{\partial T} = -\Delta S_R^o = -\frac{\Delta H_R^o}{T} \tag{4.155}$$

From equation (4.149) and (4.155), we can see that

$$\left( \frac{\partial \ln K_P}{\partial T} \right)_P = -\frac{\Delta H_R^o}{RT^2} \tag{4.156}$$

Equation (4.156) shows the effect of temperature on the equilibrium constant when the total pressure is constant.

Because equation (4.156) is a function of temperature only, we can replace the partial differentials by total differentials and integrate the equation to obtain, for small $\Delta T$ between temperatures $T_1$ and $T_2$,

$$\frac{\ln K_{P_2}}{\ln K_{P_1}} = -\frac{\Delta H_R^o}{R}\left(\frac{1}{T_2} - \frac{1}{T_1}\right) \qquad (4.157)$$

$\Delta H_R$ is the heat of reaction.

Tables 4.6, 4.7, 4.8 and 4.9, in the Appendix, list experimental and industrially used values of equilibrium constants, heats of formation and reaction, at the tenmperatures listed, for some compounds and reactions.

### 4.16.4: Calculations involving Heats of Reaction, Formation, Combustion etc.

Many calculations involving the different enthalpies of transition and reaction are based on the laws of thermochemistry which, in itself, is based on the first law of thermodynamics. The laws of thermochemistry are

a) The Law of Lavoisier and Laplace (1780 AD) and
b) The Law of Constant Heat Summation (G. Hess, 1840 AD.)

The Law of Lavoisier and Laplace (1780 AD)

At a given temperature and pressure, the amount of energy required to decompose a chemical compound is, precisely, equal to that involved in the formation of the compound from its elements.

The Law of Constant Heat Summation (G. Hess, 1840 AD)

The net heat evolved or absorbed in a chemical process is the same whether the reaction takes place in one or several steps.

The usefulness of the law of Hess is that it enables the calculation of the heats of formation of compounds from other reactions not involving direct formation of the compound from its elements. The law of Lavoisier and Laplace provides the justiflcation for the algebraic manipulation required in the use of the law of Hess.

**Illustrative Example 4.28: Calculating the Heat of Reaction in the Gaseous Phase**

Calculate the heat of reaction, at 298 K for the reaction
$$C_2H_6 + H_2 = 2CH_4 \tag{1}$$
given that the heats of formation, $\Delta H_f^{298}$ are, for $C_2H_6$, -84.459 kJ/kmol and for $CH_4$, - 74,729 kJ/kmol.

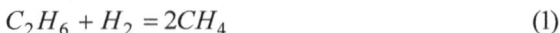

**Answer**

Since
$$\Delta H_R^{298} = \sum_{j\ products} \Delta H_f^{298} - \sum_{i\ reactants} \Delta H_f^{298} \tag{2}$$

Then
$$\Delta H_R^{298} = 2x(-74,729)-(-84,459+0)=-64,999\ kJ\ /\ kmol \quad Ans.$$

**Illustrative Example 4.29: Calculating the Temperature Dependence of the Heat of Reaction**

The heat of reaction, at 25 C, for the reaction
$$C_6H_6(g)+3H_2(g)=C_6H_{12}(g)$$
is given as $\Delta H_R^{298}$ = -206 kJ/mol. Calculate the heat of this reaction at 125 C for the reaction given that for $C_6H_6$ (g): Cp = 11.7 + 0.247 T, J/mol.K; for $C_6H_{12}$ (g): Cp = 10.9 + 0.402 T, J/mol.K; and for $H_2$ (g): Cp = 28.9 J/moLK.

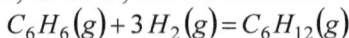

**Answer**

For the reactants, the sum of heat capacities
$$\sum_{2\ reactants} Cp = 3x28.9+11.7+0.247T = 98.4+0.247T, J/mol.K \tag{1}$$
For the product, the sum of heat capacities
$$\sum_{1\ product} Cp = 10.9+0.402T,\ J/mol.K \tag{2}$$

For the reaction
$$\Delta Cp = \sum_{j\ products} Cp_j - \sum_{i\ reactants} Cp_i = 10.9+0.402T-98.4-0.247T$$
$$= -87.5 + 0.155T, J/mol.K \tag{3}$$
Since T = 125 C + 273 = 398 K, substitution of (3) into the Kirchoff's

182

equation (4.106), we get

$$\Delta H_R^{398} = \Delta H_R^{298} + \int_{298}^{398} \Delta Cp\, dT = -206,000 + \int_{298}^{398}\left(-87.5 + 0.155\,T\right)dT$$

$$= -206,000 - 87.5\left(398 - 298\right) + 0.155\,x\,\frac{398^2 - 298^2}{2}$$

$$= -206,000 - 8750 + 5394 = -209,356\,J\,/\,mol \quad Ans.$$

It is not difficult to see that, for complicated expressions for the heat capacity, this calculation can get quite tedious, unless one is using the digital computer. Simpler methods are preferred in industrial practice.

In one of these, that is published by Chemetron Corp, USA, for customer use, the enthalpy and the logarithm of the equilibrium constant, as functions of temperature, are tabulated for the desired elements and compounds (see Appendices VIII, IX, X, XI).

The enthalpy is written as $H°-H°_f+\Delta H°_f$ where $H°-H°_f$ represents the enthalpy of the compound referred to 0 K and $\Delta H°_f$ represents the enthalpy of formation from its elements at 0 K. For elements, $\Delta H°_f$ is zero and the enthalpy is simply $H°-H°_f$.

One advantage of this method is that, since the reference state of all materials is at 0 K, the calculation of the heat of reaction, at any temperature, is obtained, easily, as the algebraic difference of enthalpies of the reactants and products of the reaction as written.

Another advantage is that, even though reactants and products come into the reaction at different temperatures, the heat of reaction at any desired temperature is, automatically, calculated using the tabular enthalpy values of the reactants and products at their entry and exit temperatures and not, necessarily, at the desired temperature.

One disadvantage of the method is that the data may not be available in this form for particular elements and compounds of interest.

**Illustrative Example 4.30: Calculating the Temperature Dependence of the Heat of Reaction**

Calculate the heat of reaction between nitrogen and hydrogen at 538 C to produce ammonia.

**Answer**

From Appendices IX and XI, the values of $H^o - H_f^o + \Delta H_f^o$ are obtained as follows $N_2$ (g), 24,079 J/mol; $H_2$ (g), 23,509 J/mol; $NH_3$ (g), -6,233 J/mol . Since the reaction, as written, is

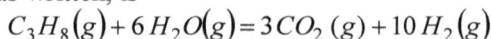

$$\tfrac{1}{2} N_2(g) + \tfrac{3}{2} H_2(g) = NH_3(g)$$

$$\Delta H_R^{538} = -6,233 - \left( \frac{24,079}{2} + \frac{3 \times 23,509}{2} \right) = -53,536 \, J \, / \, kmol \quad Ans.$$

**Illustrative Example 4.31: Calculating the Temperature Dependence of the Heat of Reaction**

Calculate the heat of reaction for the process in which propane, at 260 C, and steam, at 149 C, are reacted to produce carbon dioxide and hydrogen at 38 C.

**Answer**

The reaction, as written, is

$$C_3H_8(g) + 6 H_2O(g) = 3 CO_2(g) + 10 H_2(g)$$

From Appendices IX and XI, the values of $H^o - H_f^o + \Delta H_f^o$ for these reactants and products are obtained as follows

| Compound | Temperature, C | $H^o - H_f^o + \Delta H_f^o$, J/mol |
|---|---|---|
| Propane | 260 | -44,078 |
| Steam | 149 | -224,971 |
| Carbon Dioxide | 38 | -383,578 |
| Hydrogen | 38 | +8,844 |

Thus

$$\Delta H_R^{38} = (3x - 383,578 + 10 \times 8,844) - (-44,078 + 6x - 224,971)$$
$$= +331,610 \, J \, / \, kmol \quad Ans.$$

**Illustrative Example 4.32: Calculating the Heat of Formation in Solution**

Calculate the heat of formation for the process in which HCl forms an aqueous solution containing 8 mols of water per mol of HCl.

184

**Answer**                                    -

The heat of formation can be obtained from tables as $\Delta H_f^{298}$= -92,378 J/mol. The integral heat of solution for HCl (n = 8) is given, from Fig. 4.8 as $\Delta H_{IHS}^{298}$ = - 66,992 J/mol (1 kCal/kg-mole = 4.187 kJ/kg-mol).

Since

$$\tfrac{1}{2}H_2(g)+\tfrac{1}{2}Cl_2(g)=HCl(g) \quad with \quad \Delta H_{fg}^{298}=-92,378\,J/mol \qquad (1)$$

and

$$HCl(g)+8H_2O(l)=HCl(n=8) \quad with \quad \Delta H_{IHS}^{298}=-66,992\,J/mol \qquad (2)$$

Adding (1) and (2)

$$\tfrac{1}{2}H_2(g)+\tfrac{1}{2}Cl_2(g)+8H_2O(l)=HCl(n=8) \qquad (3)$$

$$with \quad \Delta H_{fs}^{298}=\Delta H_{fg}^{298}+\Delta H_{IHS}^{298}=-92,378-66,992=-159,370\,J/mol \quad Ans$$

**Illustrative Example 4.33: Calculating the Heat of Dilution**

Calculate the heat of diluting a solution of HCl containing 5 mols of water per mol of HCl to one containing 14 mols of water per mol of HCl.

**Answer**

Since                        $HCl\ (n=5)+9\,H_2O(l)=HCl(n=14)$ \qquad (1)

From Fig. 4.8, for HCl (n = 14), $\Delta H_{IHS}^{298}$ = -69,086 J/mol and for HCl (n = 5), $\Delta H_{IHS}^{298}$ = -62,805 J/mol, then

$$\Delta H_{HD}^{298}=\Delta H_{IHS14}^{298}-\Delta H_{IHS5}^{298}=-69,086-(-62,805)=-6,281\,J/mol \quad Ans$$

**Illustrative Example 4.34: Calculating the Heat of Solution of Hydrates**

Calculate the standard heat of solution when $Na_2CO_3.10H_2O$ is dissolved in water to form a solution containing 20 mols of water per mol of $Na_2CO_3$

**Answer**

The heats of formation are obtained from Tables as follows:

185

| Compound | $\Delta H_f^{298}$ , J/mol |
|---|---|
| $Na_2CO_3$ (c ) | -1,131,746 |
| $H_2O$ (l) | -286,045 |
| $Na_2CO_3.10H_2O$ (c ) | -4,084,837 |

For the reaction

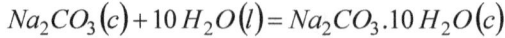

$$Na_2CO_3(c) + 10\,H_2O(l) = Na_2CO_3.10\,H_2O(c)$$

the heat of hydration is

$$\Delta H_H^{298} = -4,084,837 - (-1,131,746 + 10\,x - 286,045) = -92,641\,J/mol \qquad (1)$$

From Fig 4.9, the enthalpy of $Na_2CO_3$ (n = 20) is equal to - 30,565 J/mol $Na_2CO_3$.

Hence, for the reaction
$$Na_2CO_3.10H_2O(c) + 10\,H_2O(l) = Na_2CO_3\,(n = 20)$$
The standard heat of solution is
$$\Delta H_{HS}^{298} = -30,565 - (-92,641) = +62,076\,J/mol\ Na_2CO_3 \qquad Ans.$$

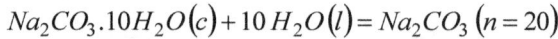

**Illustrative Example 4.35: Use of the Enthalpy – Concentration Chart**

Calculate the final temperature and concentration when 5 kg of 5 % HCl at 0 C are mixed, adiabatically, with 12 kg of 20 % HCl at 60 C.

**Answer**

The enthalpy of mixing, using enthalpy-concentration chart, is given from equation (4.116) by
$$H_S = x_1 H_1 + x_2 H_2 + x_2 \Delta H_{IHS2} \qquad from\,(4.116)$$
From Fig. 4.10, the enthalpy of 5% HCl at 0 C is -93.04 kJ/kg and that for 20 % HCl at 60 C is -186.08 kJ/kg (1 BTU per lb = 2.326 kJ/kg). Because the process is adiabatic, $\Delta H_{IHS2} = 0$  Hence
$$H_S = x_1 H_1 + x_2 H_2 + x_2 \Delta H_{IHS2}$$
$$= \frac{5}{5+12} x - 93.04 + \frac{12}{5+12} x - 186.08 + \frac{12}{5+12} x\,0 = -158.72\,kJ/kg \qquad (1)$$
The concentration of HCl in the final mixture, $C_S$, is

186

$$C_S = \frac{0.05\,x\,5 + 0.2\,x\,12}{5 + 12} = 0.1559 \quad or\ 15.6\% \tag{2}$$

Since -158.72 kJ/kg is equivalent to 68.3 BTU/lb, the temperature in Fig 4.10 which matches this enthalpy at 15.6 % HCl is 100 F which is 37.8 C. Ans

## 4.17:    Other Methods of Predicting Phase. and Chemical Equilibrium

The increasing availability of greater computational power has made it possible to combine methods of classical thermodynamics with those of statistical mechanics to be able to predict fluid phase (Winnick, 1975), and chemical (Benson, 1980), equilibrium. This is to preferred by most commercial organizations because of possible savings in cost of experimental methods.

### 4.17.1: Predicting Fluid Phase Equilibrium

The methods used are based on

- Statistical mechanics;
- The perturbation approach
- The use of models
- The use of equations of state and
- The group contribution approach.

The treatment of these methods given here is, largely, due to Winnick (1975) and Benson (1980) to which reference should be made for more detail.

Statistical Mechanics

The Helmholz energy, A, consists of two parts, namely, that which is a function of the individual, isolated, molecules themselves and that which depends on the interaction between molecules. The first part is estimated for an ideal gas. The second part, $A_C$ is the configurational contribution, given by

$$A_C = -kT \ln Z \tag{4.158}$$

where Z is the configuration integral over all spatial configurations in the system. Each configuration is weighted by $e^{U/kT}$ where U is the total potential energy of that configuration and k is the Boltzman constant.

187

When Z can be evaluated, the usual thermodynamic equilibrium properties (pressure, P, internal energy, U, and entropy, S) of the configuration can thus be obtained as

$$P = \left(\frac{\partial A_C}{\partial V}\right)_{T,n} \; ; \; U = -T^2 \left(\frac{\partial A_C/T}{\partial T}\right)_{V,n} \; ; \; S = \left(\frac{\partial A_C}{\partial T}\right)_{V,n} \qquad (4.159)$$

This method is said to be useful, only for simple fluids such as Argon and artificial (computer) liquids.

## The Perturbation Approach

The Heimholz free energy of the system is treated as a perturbation from that of a system of hard spheres in repulsion, whose properties can be expressed, by simulation, in analytic form. The perturbation is based on the attractive forces of the molecules and on an effective hard sphere diameter as a function of temperature.

When mixtures are treated, the hard sphere model is that of a mixture of hard spheres of different sizes. The method has been used to determine Henry's law constants, vapour - liquid equilibrium and transport properties for relatively, simple molecules such as argon in $H_2O$/MeOH solvents, argon in n-Pentane, etc. The method is said to be a very promising approach for the evaluation of partition functions of real fluids.

## The Use of Models

The use of models is one of the two most important industrial methods for predicting vapour-liquid equilibrium. It is based on the notion that the liquid phase can be described by structure models to simplify the configurational integral. The theory of Eyring and the cell theory of Winnick and Prausnitz have been used and result in partition functions whose three parameters, for each pure component, can be evaluated from liquid density and vapour pressure data. The additional binary parameter, $k_{ij}$, required for mixtures, is determined, empirically. Once parameters are determined, they can be used to predict multicomponent equilibria. For example, the excess Gibb's energy, $G^E$, will be given by

$$G^E = RT\left(X_1 \ln\gamma_1 + X_2 \ln\gamma_2\right) \quad or \quad \ln\gamma_1 = \frac{1}{RT}\left(\frac{\partial G^E}{\partial n}\right)_{T,P,n_2} \qquad (4.160)$$

where X is the mole fraction and $\gamma$ is the activity coefficient.

## The Corresponding States Approach

This is, essentially, an application of dimensional analysis to the
configurational portion of the partition function. The intended benefit is
that the configurational, thermodynamic properties of fluids can be
expressed as universal functions in dimensionless groups. For any gas,
for example,

$$Z = f\left(\frac{V}{V_C}, \frac{T}{T_C}\right) \tag{4.161}$$

is an application of the method where $Z$ is the compressibilty factor and
$T/T_C$, $V/V_C$ are dimensionless groups.

For mixtures, rules are developed to determine their characteristic
properties in terms of those of pure components so that these properties
can be used in the corresponding states equations. For example, for a
mixture of $i$ components

$$T_C^1 = \sum_1 x_i T_{C_i}; \quad P_C^1 = \sum_1 x_i P_{C_i}; \quad V_C^1 = \sum_1 x_i V_{C_i}; \quad W_C^1 = \sum_1 x_i w_i \tag{4.162}$$

where $w_i$ is the accentric factor.

To allow for unlike interactions between molecules, for example,

$$T_{C_{ij}} = \left(T_{C_i}.T_{C_j}\right)^{\frac{1}{2}}\left(1 - k_{ij}\right); \quad V_{C_{ij}} = \left(\frac{V_{C_{ij}}^{\frac{1}{3}} + V_{C_{ji}}^{\frac{1}{3}}}{2}\right) \tag{4.163}$$

so that

$$T_C^1 = \sum_{i=1}\sum_{j=1} x_i x_j T_{C_{ij}}; \quad V_C^1 = \sum_{i=1}\sum_{j=1} x_i x_j V_{C_{ij}} \tag{4.164}$$

This approach is said to have broad application and good accuracy but
to involve numerous and lengthy iterations which do not converge in
some regions. It, also, excludes polar and quantum fluids.

## The Use of Equations of State

This is the other of the two most important industrial methods for
predicting vapour-liquid equilibrium. In this approach, the liquid and
vapour phases are described with the same equation even though the
liquid phase may not be ideal (other methods use different equations or
assumptions.for liquid and vapour phases).

The equations of state used are

1. the ideal gas equation,
2. the Redlich-Kwong equatio
3. the Benedict-Webb-Rubin (BWR) equation, and
4. the parametric equation of state.

The best known equation, based on the ideal gas equation, is the virial equation for gases

$$\frac{PV}{RT} = 1 + \frac{A}{V} + \frac{B}{V^2} + \frac{C}{V^3} + \cdots \qquad (4.165)$$

This equation is useful for densities up to 75 % of the critical density (because coefficients of order higher than the second such as C, D, etc., are not available) and works quite well as the two phases become more identical (at high reduced temperatures and pressures).

A successful variant is the augmented virial equation of Bienkowski (Winnick, 1975) which is a combination of the hard sphere equation for liquids and the virial equation for gases. It is successful at high temperatures ($T_R > 1.5$) but not for liquids.

One of the more successful methods is based on the Redlich-Kwong equation of state and is credited to Chueh (Winnick. 1975). The mixing rules of Bienkowski were used in a modified Redlich-Kwong equation of state such that the parameters are evaluated, separately, for the liquid and vapour phases and to be independent of temperature.

Phase equilibrium calculations from the equation of state based on fugacity coefficients are, however, said to be more accurate than those estimated from Bienkowski's method. Thus the fugacities of the vapour and liquid phases are given as

$$\frac{f_i^v}{y_i\,P} = \int_{\infty}^{V}\left(\frac{RT}{V} - \left.\frac{\partial P}{\partial n_i}\right|_{T,V,n_j}\right) dV - RT\ln Z_M \qquad (4.166)$$

$$f_i^L(P) = f_i^L(P_o)\exp\int_{P_o}^{P}\left(\frac{V_i}{RT}\right)^L dP \qquad (4.167)$$

where $f_i^v$ = vapour phase fugacity, $f_i^L$ = liquid phase fugacity, $y_i$ = vapour phase mole fraction, $V_i$ = the partial molar volume, and $Z_M$ = compressibility factor for the mixture.

190

The equilibrium condition is

$$f_i^v y_i P_T = f_i^L x_i P_T \tag{4.168}$$

where $y_i$ and $x_i$ are the mole fractions in the vapour (v) and liquid (L) phases, respectively, $P_T$ is the total pressure and $f_i$ is the fugacity for component i, given by

$$\ln \frac{f_i}{P_i} = \int_{P=0}^{P} \left( \frac{V}{RT} - \frac{1}{P} \right) d P \tag{4.169}$$

Once the equation of state is known for both phases, it is used to solve for fugacity in equation (4.169) which, in turn, is used to solve equation (4.168).

When it is not possible to use the equation of state, the activity coefficient approach, an elaboration of the models approach, gives, at phase equilibrium,

$$f_i^v y_i P_T = \gamma_i X_i f_i^o \exp\left( \frac{1}{RT} \int_{P_{ref}}^{P_T} V_i d P \right) \tag{4.170}$$

where $\gamma_i$ is the activity coefficient in the liquid phase of component i.

Activity coefficients may be predicted using the Wilson equation for totally miscible solutions, or the NRTL (Non-Random Two Liquid) equation for solutions exhibiting only partial miscibility. These are the Wilson's equations

$$\ln \gamma_1 = -\ln(x_1 + A x_2) + x_2 \left( \frac{A}{x_1 + A x_2} - \frac{B}{B x_1 + x_2} \right) \tag{4.171}$$

$$\ln \gamma_2 = -\ln(B x_1 + x_2) - x_1 \left( \frac{A}{x_1 + A x_2} - \frac{B}{B x_1 + x_2} \right) \tag{4.172}$$

and the NRTL equation

$$\ln \gamma_1 = \frac{1}{RT} \left( \frac{\partial G^E}{\partial n_i} \right)_{T,P,n_2} \tag{4.173}$$

where $G^E$ is the excess Gibb' s free energy.

Other less common methods of determining activity coefficients have been described by Winnick (1975). Buck (1984) considered six variations of equation (4.168) for the system, ethanol-water at 1.01 to 20.7 bar and concluded that the version, given by equation (4.174) below, was the only acceptable one for the prediction of data for design purposes.

$$f_i^v y_i P_T = \gamma_i x_i P_i f_i^L \exp\left(\frac{V_i (P_T - P_i)}{RT}\right) \tag{4.174}$$

where $P_i$ is the vapour pressure of pure liquid component, *i*. V may or may not be a function of temperature

The most successful modification of the Redlich-Kwong equation is that due to Soave (Winnick, 1975) and is based on estimating the parameter $\alpha$ as a temperature dependent parameter using pure component vapour pressure data. The results for non - polar fluids and mixtures of such fluids are good and without the use of experimental mixture data.

A two parameter equation of state (like the Redlich-Kwong, van der Waal's), which is a combination of the hard sphere model and the Redlich-Kwong attractive term, has been tested by Carnahan and Starting (Winnick, 1975). The two parameters describe molecular size and intermolecular attraction.

The equation presents properties, such as density, enthalpy and fugacity; of polar non spherical molecules and non-polar spherical molecules better than the original Redlich-Kwong equation. The constants of this equation are evaluated from critical constants, as in the van der Waal's equation.

When sufficient data are available, the Benedict-Webb-Rubin (BWR) equation is preferred. It is given as

$$P = RT\rho + \left(BRT - A - \frac{C}{T^2}\right)\rho^2 + (bRT - a)\rho^3$$

$$+ a\alpha\rho^6 + \frac{c\rho^3\left(1 + \gamma\rho^2\right)}{T^2} \tag{4.175}$$

w'here A, B, C, a. b, c, $\alpha$ and $\gamma$ are constants.

Several modifications have been made by many workers in the field to obtain better mixture property correlation. A major consideration is the method used to obtain the BWR coefficients either in terms of reduced properties and the accentric factor or in terms of parameters for binary pairs or in terms of adjustable parameters which, in mixtures, are a function of composition.

Polymer solutions are better described by the Flory-Huggins equation

192

$$\frac{\mu_1 - \mu_1^o}{RT} = \ln(1 - \varphi_2) + \left(1 - \frac{1}{r}\right)\varphi_2 + x \cdot \varphi_2^2 \qquad (4.176)$$

where $\mu_1^o$, $\mu_1$ are the chemical potentials of pure solvent out of and in solution, respectively, $\varphi_2$ is the volume fraction of polymer, r is the molar volume ratio of polymer to solvent and x is an empirical constant, dependent on composition.

The Flory-Huggins equation is, also, useful in the calculation of the activity coefficient of polymers as a function of concentration, solubility of crystalline polymers, melting point depression, swelling of cross-linked polymers in solvents and the miscibility boundaries of polymer solutions.

### 4.17.2: Predicting Chemical Equilibrium

The information, usually, sought with regard to chemical equilibrium for any chemical reaction is the value of its equilibrium constant. For a reaction such as

$$aA + bB \leftrightarrow rR + sS + \ldots\ldots \qquad (4.177)$$

Since

$$-\Delta G_R^o = RT \ln K_P \qquad \qquad from \ (4.144)$$

and

$$\Delta G_R^o = \Delta H_R^o - T\Delta S_R^o \qquad \qquad from \ (4.152)$$

knowledge of $\Delta H^o{}_R$ and $\Delta S^o{}_R$ at the standard temperature enables $\Delta G^o{}_R$ and hence $K_P$ to be determined. The value of $K_P$ at any other temperature can be estimated using the Kirchoff's equation if the molar heat capacities of the species are known.

The molar heat capacity and the standard entropy terms, but not the standard heat of formation, $\Delta H^o{}_f$, can be predicted for molecules, radicals and ions with good accuracy from statistical mechanics and empirical knowledge of their molecular structure. To predict $\Delta H^o{}_{f,,}$ additivity laws (Benson, 1980), whose discussion is beyond the scope of this book, are used

### 4.17.3: Predicting Chemical Reaction Rates

When a chemical reaction proceeds by way of an activated complex, it is possible to predict its reaction rate from thermodynamics. Suppose the reaction is

193

$$A + B \leftrightarrow M \rightarrow C \tag{4.178}$$

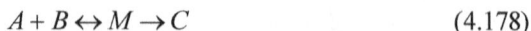

where M is the activated complex and C is the product. The reaction rate is determined, first by the equilibrium between M on one hand and A and B on the other, controlled by the decomposition of M to C. Hence

$$K = \frac{a_M}{a_A \cdot a_B} = \frac{\gamma_M \cdot [M]}{\gamma_A \cdot [A] \cdot \gamma_B \cdot [B]} \tag{4.179}$$

where $a_M$, $a_A$ and $a_B$ are the activities of M, A, and B and $\gamma_M$, $\gamma_A$ and $\gamma_B$ are their coefficients.

The apparent rate of reaction, $r_{app}$, is given by

$$r_{app} = cons \tan t . [M] = cons \tan t . K . \frac{\gamma_A \cdot \gamma_B}{\gamma_M} . [A].[B] \tag{4.180}$$

Consideration of solvent effects indicates that they can be enormous and that the rate of reaction in one solvent can be determined from that in another solvent (Winnick, 1975). This leads to the determination of activity coefficients from regular solution theory. Thus, taking $V_i$ to be the molar volume of component i, $\Delta H_V$ its enthalpy of vaporisation and $\delta_o$ its average solubility parameter (taken as that of the solvent), we get

$$\ln \gamma_i = \frac{V_i}{RT} (\delta_i - \delta_o)^2 \tag{4.181}$$

where

$$\delta_i = \left( \frac{\Delta H_{V_i}}{V_i} \right)^{1/2} \tag{4.182}$$

The problem is to determine $\gamma_M$ and $V_M$ for the activated complex. A more empirical approach shows that the rate of reaction, r, in a substitute solvent, and that in some chosen standard solvent, $r_o$, are related as

$$\log \left( \frac{r}{r_o} \right) = \lambda \tau \tag{4.183}$$

where $\lambda$ is specific to a given reaction and $\tau$ is specific to a given solvent. This is reported to give good results for highly polar reactions (Winnick, 1975).

## *References For Chapter Four*

1    Benson, S.W (1980); *Predicting Chemical Reactivity*, ChemTech. No.2; pp 121-126; Am. Chem. Soc., Wash. D.C., U.S.A. .

2    Buck, E (1984); *Applying Phase Equilibrium Thermodynamics*; ChemTech;.No.8; pp 570-575; Am. Chem. Soc., Wash. D.C., U.S.A.

3    Girdler Catalysts; *Physical and Thermodynamic Properties of Elements and Compounds*; Chemetron Corp., Louisville, KY, U.S.A (1969).

4    Green, D. W. (Editor): *Perry's Chemical Engineers Handbook*, 6th Edition; Chapter 4; McGraw-Hill Book Co.; N. Y., U.S.A. (1984)

5    Hougen, O. A, Watson, K. M., and Ragatz, R A; *Chemical Process Principles, Part 1: Material and Energy Balances*. 2nd. Edition; Chapter 9; John Wiley & Sons Inc., N. Y., U.S.A(1954)

6    Prausnitz, J. M; *Molecular Thermodynamics of Fluid Phase Equilibria*; Chapters 3 and 5; Prentice Hall, N.J., U.S.A (1969)

7    Schmidt, A. X and List, H. L.; *Material and Energy Balances*; Chapter 4; Prentice Hall; N.J., U.S.A. (1962)

8    Sinnott, R K.; *Chemical Engineerihg, Vol. 6: J M.Coulson and J. F. Richardson - An Introduction to Chemical Engineering Design*; Chapters 2 and 3; Pergamon Press, Oxford, UK (1983).

9    Winnick. J. (1975); *The State of Affairs in Thermo, Part 1*; ChemTech, March; pp 177-185; Am. Chem Soc., Wash. D.C., U.S.A.

10   Winnick, J. (1975); *The State of Affairs in Thermo, Part 2*; ChemTech, December; pp 756-762; Am. Chem. Soc., Wash. D.C., U.S.A

11   Zemansky, M. W.; *Heat and Thermodynamics*. 4th Edition; Int'l Student Edition; Chapter 3; McGraw Hill Book. Co. Inc., N. Y., U.S.A. (1957).

12   Hanna R; *HandsDown Software HDPsyChart Psychrometric Analysis Program*, (March, 2010)

13   Stoy M. A and Fricke A. L.; (1994); *Enthalpy Concentration Relations for Black Liquor*, Tappi Journal, Vol. 77 No. 9; pp 103-110

# CHAPTER FIVE
# MATERIAL AND ENERGY BALANCES

## 5.1:  Basic Concepts

As a subject develops and becomes complex in order to deal with an increasing number and variety of problems presented to it, its jargon and terminology base expands proportionately. Their purpose is to provide precision and clarity in discussions, in design and analysis and in operating procedures. Some of the basic and essential definitions of material and energy balances, as used in process design and development, are, therefore, presented below.

### 5.1.1:  Definition of System Boundary

The system boundary defines the part of a process being considered. Judicious definition and use of the system boundary can simplify calculations considerably. In Fig. 5.1, for example, boundaries A, B or C can be defined and each will yield varying degrees of calculation complexity or usefulness of results.

**Fig. 5.1:  Definitions of System Boundaries**

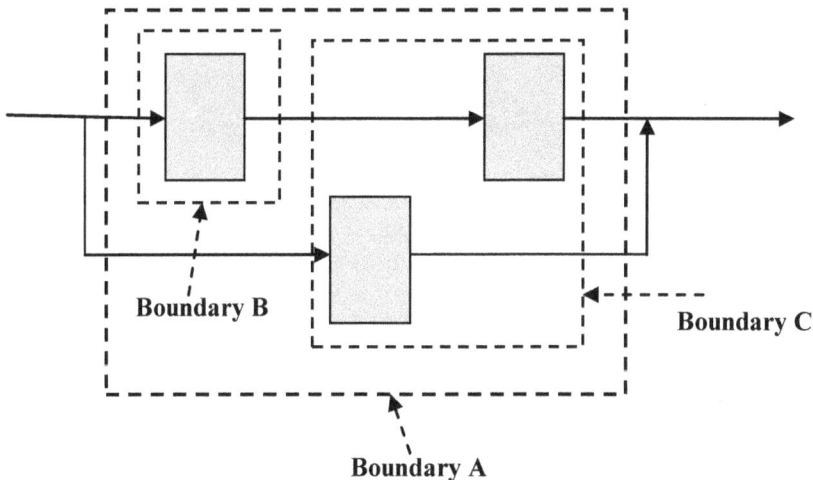

### 5.1.2:  Defining the Basis of Calculations

A basis of calculation establishes the numerical reference quantity or unit of a product, feed or stream, on which the calculation is based.

197

Such a basis of calculation can be, for example, 1000 tonnes of a substance or 50 tons/h of some material or 50 hours of operation or even 5 cycles of operation, etc. Again, judicious use, not only simplifies calculations, but also makes the results much more useful for process evaluation or for decision making. There is no particular rule for choosing the basis of a calculation except common sense which dictates that it should be tied to the most important or expensive or commonly occurring component, product or time.

### 5.1.3: Reacting and Non - Reacting Systems

Reacting systems, because they involve the consumption and production of materials, influence the scope and complexity of material balance processes. Typical reacting systems include chemical, combustion, electrical and biochemical reactions. Non-reacting systems affect the scope of the mass balance. They are, usually, pure inventory systems, such as tank or material storage or assembly/separation processes such as dissolution, mixing, evaporation, distillation, etc.

### 5.1.4: Performance and Design Parameters

These provide the indices of performance or the objectives by means of which the process can be operated, described, evaluated or compared to another process. Such parameters include stoichiometry, conversion, yield, excess reagent, tie component, etc.

### 5.1.5: Methods for Material and Energy Balances

The many calculation procedures used in carrying out material balances may be classified into three categories. The first and earliest methods may be grouped under the arithmetic methods because they involved tabular, direct and manual, component by component calculations.

A modern development of the arithmetic method is the process synthesis method of Rudd et al (1978). It is not, strictly speaking, a purely, material balance procedure since out of its four major steps,some of are concerned with economic feasibility, some with technical feasibility, although all are tackled using arithmetical, material balance procedures.

The main calculation steps are

198

a) Reaction Path Synthesis
b) Material Flow Synthesis
c) Separation Task Selection and
d) Task Integration.

Systematic application of engineering judgement in allocating species to processes and units, in material balances, knowledge of separation technology and the engineering of process systems are said to lead to optimal process design.  More details can be found in the book by the authors.

The next group of methods are the algebraic methods in the sense that algebraic symbols and procedures are used to generalise and systematise the material and energy balances. The benefit was that, not only would a particular problem be solved with the equations developed, but an optimum solution could, also, be found. A whole class of problems of similar nature could be similarly addressed without having to start afresh, as in the arithmetic methods.

These methods make use of algebraic symbols and procedures to develop and solve the material and energy balance equations around, not only each sub-system of the process but also, the whole plant. Even though the method can be, mathematically, complex, it has the advantage of being able to handle very complex processes and is, easily, amenable to computerisation.

The essence of the method is to develop consistent labels for every process variable in the system as it concerns whatever process unit it becomes involved in throughout the system. Material and energy balances across each process unit unit are collated such that this consistent labeling of variables makes it possible to have a system of algebraic equations which describe the entire system or portions of the system which may be of interest. These equations are then solved subject to the constraints in the system. This is the method of constraints.

While anybody can develop his or her own algebraic equations for his or her process as well as the solution algorithms, it is sometimes cheaper to look for generalised procedures or software rather than develop one's own. Many of such procedures and software, mostly proprietary, have been developed by commercial process design and development organizations and were listed in an earlier chapter. Only those that have entered the public domain can be discussed here. One of these is the the the split fraction method.

Even more modern, than the relatively simple algebra that are utilised in these two methods, are mathematical modelling and simulation techniques. The discussion of two algebraic methods, however, provides understanding and insight into the basis and development of more modern methods as well as providing, still, useful tools for solving many real problems. Both methods rely on a modular approach to build up a mathematical solution for the whole process.

The third group of methods, modeling, developed from the sophistication reached by the algebraic methods. Each algebraic set of equations describing a process or set of processes was, in fact, regarded as describing, or giving the ability to predict, the behavior of the process.

Modeling has evolved, recently, into simulation in which mathematical relationships, not necessarily arising from direct physical relationship to process phenomena, represent the process as well as the physically based equations. Their main advantage is that they save expensive experimentation and are more conducive to process control. Their main disadvantage is that they are unable to account for unexpected physical phenomena which may turn up in the process.

The complexity of the mathematics involved increases, as would be expected, as one goes from using the arithmetic, through the algebraic to the modeling and simulation methods. This chapter attempts tp provide the basis for the development of any of the methods by highlighting the fundamental elements and basis for material and energy balances in a process, or group of process, units.

### 5.1.6: Inventory, Assembly and Separation Operations

These are operations in which only physical, but not chemical, changes take place. Inventory operations involve changes of mass or volume in a fixed frame of reference, such as filling or emptying of a tank or container vessel. Assembly operations involve mixing, blending or compounding processes in which no chemical reactions occur. Separation operations involve processes in which the components of a mixture are separated by means of physical processes only.

200

Symbolically, we can represent any system we wish to analyse as shown below

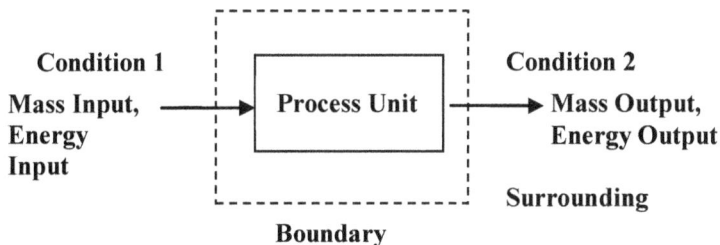

**Boundary**

If we represent the input side of the process unit as condition 1 and the output side as condition 2, we can designate the operating interval, $\Delta t$, as

$$\Delta t = t_2 - t_1 \tag{5.1}$$

By the law of conservation of mass,

$$m_2 + I_2 = m_1 + I_1 \tag{5.2}$$

Similarly, by the law of conservation of energy,

$$E_2 + EI_2 = E_1 + EI_1 \tag{5.3}$$

where, for mass, I is the inventory and $m_1$ and $m_2$ are the mass inputs and outputs in the system, while EI, $E_1$ and $E_2$ are the corresponding energy equivalents. Hence

$$m_1 = m_2 + I_2 - I_1 = m_2 + \Delta I \tag{5.4}$$

$$E_1 = E_2 + EI_2 - EI_1 = E_2 + \Delta EI \tag{5.5}$$

where $\Delta I$ and $\Delta EI$ are the changes of inventory of mass and energy in terms of the final minus the initial value. If any of $\Delta I$ or $\Delta EI$ is positive, we say that there has been accumulation in the system while if $\Delta I$ or $\Delta EI$ is negative, there is depletion.

Closed and Open Systems

The definitions of closed and open systems follow those of thermodynamics. A closed system exchanges only energy but not mass with the surroundings. That is

$$m_1 = 0; \quad m_2 = 0; \quad \Delta I = 0 \tag{5.6}$$

An open system exchanges both energy and mass with the surroundings. That is

$$m_1 = m_2 + I_2 - I_1 = m_2 + \Delta I \qquad \textit{from} \quad (5.4)$$

$$E_1 = E_2 + EI_2 - EI_1 = E_2 + \Delta EI \qquad \textit{from} \quad (5.5)$$

At steady state,

201

$$m_1 = m_2 \qquad and \qquad \Delta I = 0 \qquad\qquad (5.7)$$

$$E_1 = E_2 \qquad and \qquad \Delta EI = 0 \qquad\qquad (5.8)$$

### 5.1.7:  Basic Definitions of Process Units and Operations

Unit Operation

This is a chemical engineering process which involves only physical but not chemical changes.

Unit Process

This is a chemical engineering process which involves both chemical and physical changes.

Batch Process

This is a process which comes to an end during the operating interval, as a result of the depletion of one or more essential material components (Schmidt & List, 1962). It can be either a unit operation or a unit process and can take place in an open or closed system. Typical batch processes are the cooking of food (open system) or reaction in a bomb calorimeter (closed system).

Continuous Process

This is a process in which all essential materials are available throughout the operating interval and the termination of the process is not as a result of the depletion of one or more essential material components (Schmidt & List, 1962). Continuous processes are possible, only, in open systems and, in addition, may involve either variable or invariant flow.

Single Operation Process

This involves one single operating or process unit such as evaporation, filtration, reaction, etc. Schematically, a single operation process is shown below

**Fig. 5.2: Single Operating Process**

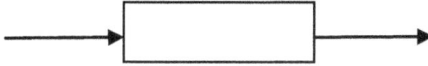

Consecutive Operation Process

A consecutive operation process involves more than one operating or process unit in series. Schematically

**Fig. 5.3: Consecutive Operating Process**

Co-current Operation Process

A co-current process is such that both streams enter and leave the process unit in the same direction. Co-current processes are associated with higher driving forces at the inlet than at the outlet. Stream properties, for both streams, tend to move towards an asymptote.

**Fig. 5.4: Cocurrent Process**

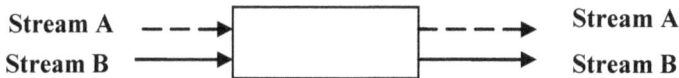

Stream A    Stream A
Stream B    Stream B

The figure, below, called an exchange diagram, illustrates the variation of the variable of interest along the length of the process unit.

When the operation is heat transfer, the variable of interest is the temperature difference while in mass transfer, this variable is the concentration difference. Co-current processes are more suited to the processing of streams and products which are sensitive, adversely, to high driving force. Co-current processing, because of decreasing driving force towards the exit, tends to require large equipment capacity.

**Fig. 5.5:  Exchange Diagram for a Cocurrent Process**

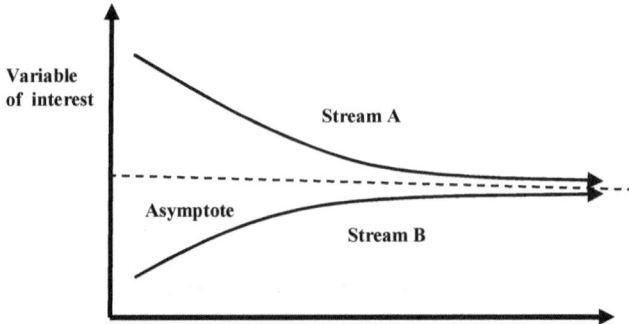

Counter - current Process     **Distance along unit**

Here, both streams enter and leave the process unit in opposite
directions.

**Fig. 5.6:  Countercurrent Process**

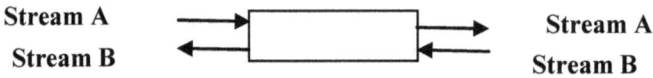

The exchange diagram, for this process, illustrated below, shows that
the driving force is much more uniform throughout the process. This
has the advantage of higher processing efficiency, thereby requiring
smaller equipment capacity than the co-current process. It is not
suitable, however, for streams and products which are sensitive to high
driving force at exit conditions.

**Fig. 5.7:  Exchange Diagram for a Countercurrent Process**

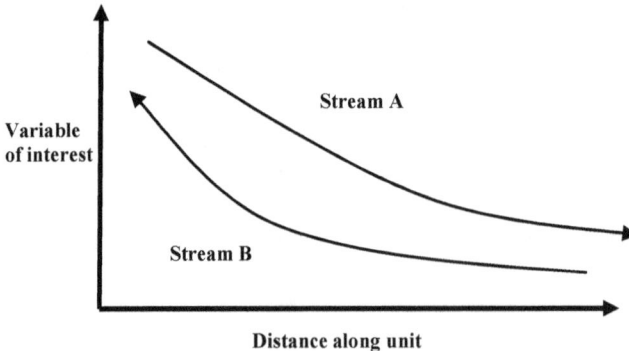

204

## Recycle Process

Recycle processes involve the reuse of materials which have passed through a process operation. Recycle (stream E in Fig. 5.8) is the emergent stream which is returned to the operating unit to be re-processed. Total recycle means that all processed material is returned for re-processing while partial recycle means that only some of the processed material is returned for re-processing.

Fresh or Net Feed (stream A in Fig.5.8) is defined as part of the material fed to the processing unit but not including recycled material. Total or Combined Feed (stream B in Fig. 5.8) is, then, the fresh or net feed plus recycle. Recycle ratio is the ratio of recycle to fresh feed in consistent units.

**Fig. 5.8:  Recycle Process**

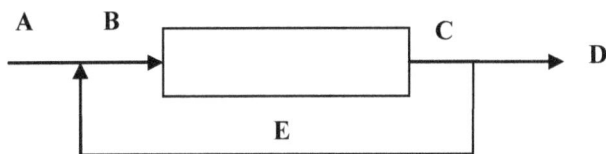

## By-Pass Process

This is the process in which entering material is split such that one part goes through the operating unit while the other by-passes it (stream F in Fig.5.9. The purpose is, either to reduce the quantity of material to be processed or to maintain uniformity of quality in the product.

**Fig. 5.9:  Bypass Process**

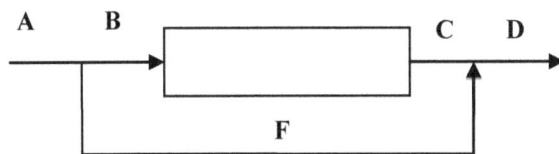

## Purge Process

This is the process in which a portion of the recycle stream is bled off in order to prevent the build-up of unwanted material in the process. Such unwanted materials are, usually, inert or do not react with other materials in the processing unit. Purge processes are important, for example, in pollution control where there is need to meet specifications on limits of the concentration of certain components in exit streams or in a given product.

**Fig. 5.10:  Purge Process**

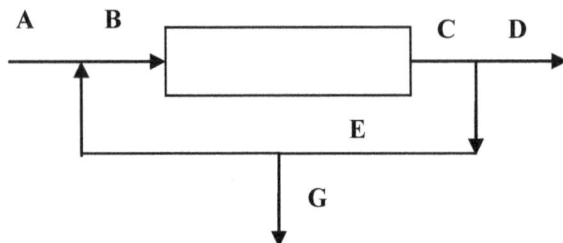

Recycle, by-pass or purge processes enable chemical process operators to meet, economically, various standards or specifications required in the market place or by government and safety regulations.

## 5.2:  Material Balances Across Individual Process Units

**Illustrative Example 5.1:  Pure Inventory Operations**

50 gallons per minute of a fluid flows into a tank from which 30 gallons per minute flows out. If the operating interval is 30 minutes, in a tank with an initial inventory of 3000 gallons, calculate the inventory at the end of the operation.

**Answer**
Input = 50 x 30  = 1500 gallons
Output = 30 x 30 = 900 gallons
From (5.2)
$$m_1 + I_1 = m_2 + I_2 = 1500 + 3000 = 900 + I_2$$
That is $I_2 = 3600$ gallons. Ans

**Illustrative Example 5.2:  Single Operating Process**

The feed to a small scale refinery distillation column consists of a mixture of propane, butane and pentane in equal weights and is at the rate of 1000 kg/h. The overhead contains 90 % propane, 7.0% butane and 3.0 % pentane and is produced at the rate of 280 kg/h. The side stream is produced at the rate of 350 kg/h and contains 20 % propane, 70 % butane and the rest pentane.  Calculate the weight of the bottoms stream and its composition.

**Answer**

Basis: 1 hour of operation

The operation may be represented as shown below. Since there is no change in inventory; for steady state operation;

$$Total\ mass\ in = Total\ mass\ out \tag{1}$$

$$F = E + S + B \tag{2}$$

where F, E, S, and B are the mass flow rates of the feed, overhead, side and bottom streams, respectively. Thus

$$B = F - E - S = 1000 - 280 - 350 = 370\ kg/h \quad Ans$$

To calculate the composition of the bottoms stream, a component balance gives for propane:

$$x_F F = x_E E + x_S S + x_B B \tag{3}$$

where $x_F$, $x_E$, $x_S$, and $x_B$ are the weight fractions of propane in the feed, overhead, side and bottom streams, respectively.

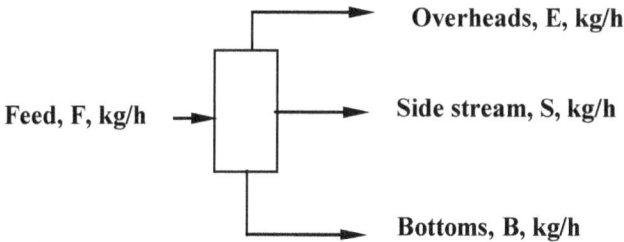

That is

$$0.333 \times 1000 = 0.9 \times 280 + 0.2 \times 350 + x_B \times 370$$

from which

$$x_B = \frac{0.333 \times 1000 - 0.9 \times 280 - 0.2 \times 350}{370} = 0.03 \quad Ans$$

For butane

$$x_F F = x_E E + x_S S + x_B B \tag{4}$$

where $x_F$, $x_E$, $x_S$, and $x_B$ are the weight fractions of butane in the feed, overhead, side and bottom streams, respectively. That is

$$0.333 \times 1000 = 0.07 \times 280 + 0.7 \times 350 + x_B \times 370$$

from which

$$x_B = \frac{0.333 \times 1000 - 0.07 \times 280 - 0.7 \times 350}{370} = 0.19 \quad Ans$$

Since

207

$$x_B(propane) + x_B(bu \tan e) + x_B(pen \tan e) = 1 \qquad (5)$$

$$x_B(pen \tan e) = 1 - x_B(propane) - x_B(bu \tan e) = 1 - 0.03 - 0.19 = 0.78 \; Ans.$$

Thus the weight of Bottoms is 370 kg/h. Its composition is 3.0 % propane, 19 % butane and 78 % pentane. Ans

## Illustrative Example 5.3:  The Countercurrent Operating Process

A benzene soaked filter cake, containing 20 % benzene and 80 % inert solids, is being freed of benzene by heating in a stream of nitrogen in a continuous countercurrent dryer. The nitrogen enters dry and removes 0.7 kg benzene per kg nitrogen as it passes through the system. The exiting solid retains 4.0 % by weight of benzene. How many kg of nitrogen would be required to pass through the dryer per kg of inert solids?

**Answer**

N₂ in, $F_{11}$ ——→ | Counter current Dryer | ——→ N₂ out, $F_{12}$

Filter cake out, $F_{21}$ ←—— | | ←—— Filter cake in, $F_{22}$

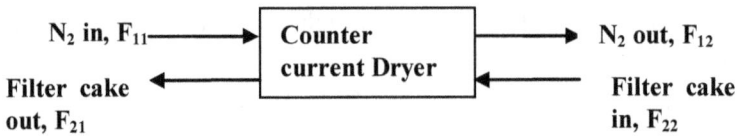

Basis: 1 kg of inert solids through the dryer

Because there are two parallel streams going through the process unit, the nomenclature, for the streams, becomes a bit more complicated. Because nitrogen is the stream we are interested in estimating the requirement, we shall label it as stream 1 and hence the benzene soaked filter cake stream as stream 2. Taking the side in which the nitrogen stream enters as side 1, we shall label its exit side as side 2. The nitrogen inlet stream, thus, becomes stream $F_{11}$, that is, stream 1 entering in side 1. Similarly, for the other streams, as shown in the diagram.

Assuming steady state, all materials entering may be estimated by a material balance as

$$F_{11} + F_{22} = F_{12} + F_{21} \qquad (1)$$

A benzene balance gives

$$x_{F_{11}}F_{11} + x_{F_{22}}F_{22} = x_{F_{12}}F_{12} + x_{F_{21}}F_{21} \qquad (2)$$

Mass fraction of benzene in the feed nitrogen stream is

$$x_{F_{11}} = 0 \; kg \; benzene \; / \; kg \; nitrogen \; stream \qquad (3)$$

208

Mass fraction of benzene in the exit nitrogen stream is

$$x_{F_{12}} = 0.7 \, kg \; benzene \, / \, kg \; nitrogen \; stream \qquad (4)$$

Since the basis of calculation is 1 kg inert solids, then mass of benzene in the filter cake feed stream is

$$= \frac{0.2}{0.8} = 0.25 \, kg \; benzene \, / \, kg \; inert \; solids \qquad (5)$$

Similarly, mass of benzene in the filter cake exit stream is

$$= \frac{0.04}{0.96} = 0.042 \, kg \; benzene \, / \, kg \; inert \; solids \qquad (6)$$

Thus the mass inputs and outputs are

$$F_{22} = 1.25 \, kg \; filter \; cake \, / \, kg \; inert \; solids$$
$$F_{21} = 1.042 \, kg \; filter \; cake \, / \, kg \; inert \; solids \qquad (7)$$
$$F_{11} = 1 \, kg \, / \, kg \; nitrogen \; stream$$
$$F_{12} = 1.7 \, kg \, / \, kg \; nitrogen \; stream$$

The mass fractions are

$$x_{F_{22}} = 0.20 \, kg \; benzene \, / \, kg \; filter \; cake$$

$$x_{F_{21}} = \frac{0.042}{1.042} = 0.04 \, kg \; benzene \, / \, kg \; filter \; cake \qquad (8)$$

Thus if $F_1$ is the rate of nitrogen per kg inert solids, then from (1) and (7)

$$F_1 + 1.25 = 1.7 \, F_1 + 1.042 \qquad (9)$$

That is

$$F_1 = \frac{0.208}{0.7} = 0.297 \, kg \; nitrogen \, / \, kg \; inert \; solids \qquad Ans.$$

From (2), (3), (4) and (7)

$$0 \, x \, F_1 + 0.20 \, x \, 1.25 = 0.7 \, x \, F_1 + 0.04 \, x \, 1.042 \qquad (10)$$

That is

$$0.20 \, x \, 1.25 - 0.04 \, x \, 1.042 = 0.7 \, F_1$$

$$F_1 = \frac{0.2083}{0.5817} = 0.298 \, kg \; nitrogen \, / \, kg \; inert \; solids \; Ans \quad (checks \,)$$

## Illustrative Example 5.4:  The Recycle Process

In a typical fractionating tower, 5000 kg/h of feed, containing propane, butane and pentane in equal proportions, is separated into an overhead product rich in propane and a bottoms product rich in butane and pentane. Part of the overhead and bottoms are recycled as reflux.
The overhead product is withdrawn at the rate of 3000 kg/h and

contains 80 % propane and 20 % butane while the bottoms product is withdrawn at the rate of 5000 kg/h containing 10 % propane, 40 % butane and 50 % pentane with 1500 kg/h recycled.

a) What is the weight of reflux from the overhead stream?
b) What is the fraction of propane in feed which goes out with the bottoms product?
c) What are the recycle ratios at the top and bottom of the fractionating tower?

**Answer**

Let us label the streams as shown in the diagram below.

Basis: 1 hour of operation

It is always better to start from the part of the problem where there is the most information available or given. Thus, from the bottoms streams, B = 5,000 kg/h, $R_2$= 1,500 kg/h so that

$$B_1 = B - R_2 = 5,000 - 1,500 = 3,500 \, kg / h \tag{1}$$

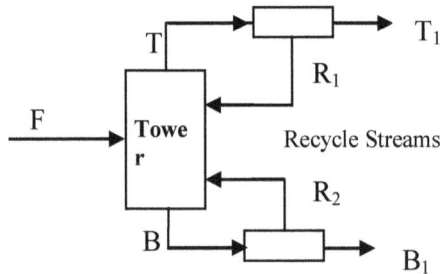

For part (a)

T was given as 3,000 kg/h, so we can say that since

$$F + R_1 + R_2 = T + B \quad or \quad R_1 = T + B - F - R_2$$
$$R_1 = 3,000 + 5,000 - 5,000 - 1,500 = 1,500 \, kg / h \qquad Ans$$

Since T = 3,000 kg/h and $R_1$ = 1,500, then

$$T_1 = T - R_1 = 3,000 - 1,500 = 1,500 \, kg / h \tag{2}$$

For part b

Since the compositions of streams $T_1$ and $R_1$ are the same for the overhead products while those of $B_1$ and $R_2$ are the same for the bottoms products, an overall propane balance will give the total amount of propane in the feed as

$$x_F F = x_{T_1} T_1 + x_{B_1} B_1 = 0.8 \times 1,500 + 0.10 \times 3,500 = 1,550 \, kg / h \qquad (3)$$

at the mass fraction of

$$x_F = \frac{1,550}{F} = \frac{1,550}{5,000} = 0.31 \qquad (4)$$

The amount of propane in the bottoms product B is

$$x_B B = 0.1 \times 5,000 = 500 \, kg / h \qquad (5)$$

Thus, from (3) and (5), the fraction of propane in the feed which goes out with the bottoms product is

$$= \frac{500}{1,550} = 0.32 \qquad \qquad Ans$$

For part (c )

Since the recycle ratio is the ratio of recycle to fresh feed, at the top of the tower, the recycle ratio is

$$\frac{R_1}{T} = \frac{1,500}{3,000} = 0.5 \qquad \qquad Ans$$

And at the bottom of the tower, it is

$$\frac{R_2}{B} = \frac{1,500}{5,000} = 0.3 \qquad \qquad Ans$$

**Illustrative Example 5.5: Bypass Process**

A waste process liquid containing 500 ppm of a toxic impurity is to be treated so that the discharge contains not more than 100 ppm. The treating process removes all but 10 ppm. What fraction of waste liquor may be bypassed?

**Answer**

The problem may be illustrated by the diagram below.

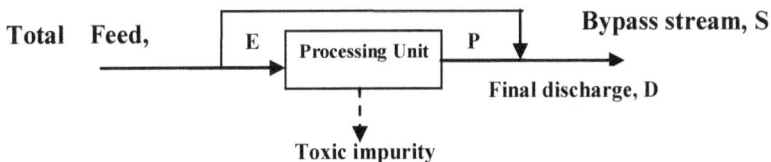

**Toxic impurity**

Basis: 1 Million Parts of Waste Liquor

Let $\alpha$ be the fraction of waste liquor bypassed.

$$Amount\ processed, E\ =\ (1-\alpha)F \qquad (1)$$

$$Amount\ bypassed, S\ =\ \alpha\ F \qquad (2)$$

Assuming steady state and no loss of carrier liquid with removed toxic impurity, material balance across the exit re-mixing unit gives

$$D\ =\ P+S = E+S \qquad (3)$$

$$x_D D = x_E E + x_S S = 10\ E + 500\ S = 100\ D \qquad (4)$$

Multiplying (3) by 10

$$10\ D\ =\ 10\ E\ +10\ S \qquad (5)$$

Subtracting (5) from (4)

$$90\ D\ =\ 490\ S \quad or \quad D = \frac{490}{90}\ S = 5.44\ S \qquad (6)$$

From (6) and (3)

$$5.44\alpha F\ =\ (1-\alpha)F + \alpha F = F \quad or \quad \alpha = \frac{1}{5.44} = 0.184\ Ans$$

## Illustrative Example 5.6:  The Consecutive Operation Process

A mixture of benzene, toluene and xylene is separated by continuous fractionation in two towers operating in series. Tower 1 produces xylene as bottoms product and a mixture of benzene and toluene as overhead products. The overhead is pumped into a holding tank and then into tower 2 where benzene is recovered as overhead and toluene as bottoms.

During a 24-hour run, 10,000 gallons of bottoms with a composition 0.7 % toluene, 99.3 % xylene by volume were recovered from tower 1 while 7,300 gallons of overhead, containing 98.5 % benzene, 1.5 % toluene and 12,000 gallons of bottoms, 1.0 % benzene, 98.5 % toluene and 0.5 % xylene, were recovered from tower 2.

The holding tank capacity is 1 gallon/mm. At start, its liquid level was 1.0 m and at the end of 24 hours, 1.5 m.

(a). What is the rate of flow from the top of tower 1?
(b). What is the volume and composition of feed during the 24-hour period?
(c ). What is the production of overhead and bottoms from tower 2 if the holding tank level remained constant during the 24-hour operating period for the same feed rate?

**Answer:**

<u>Basis</u>: 24 hour operating period

The process schematics may be represented as shown below, with the dotted lines showing the boundaries chosen for each process for analysis.

The compositions, as used in this example, are expressed with respect to the streams in which they occur. For example, benzene fraction in the feed steam $F_1$ and overhead stream $T_1$ can be expressed as $x_{F1}$ and $x_{T1}$ respectively. It is also assumed that all the streams are approximately of the same liquid density.

<u>Part (c )</u>

Start with this part of the problem (envelope D) because there is more information about tower 2. Since the holding tank level remained constant, there was no change in inventory there and hence $T_1 = F_2$. Thus, for the material balance around envelope D; assuming ideal solution

<u>Overall Balance</u>

Total feed to tower 2, $\qquad F_2 = T_2 + B_2 = 7{,}300 + 12{,}000 = 19{,}300$   (1)

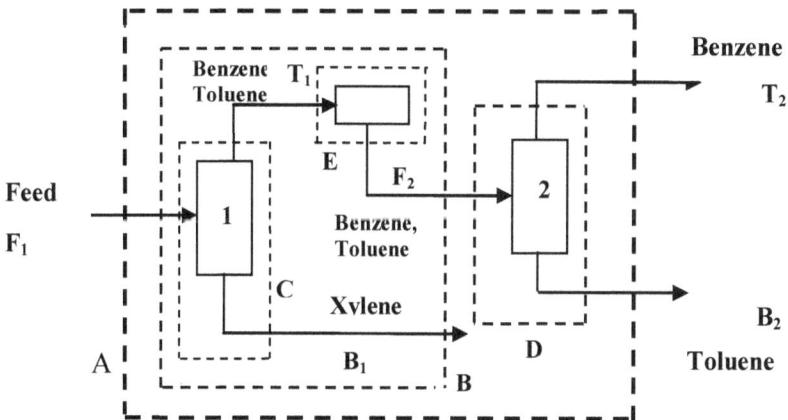

<u>Benzene Balance</u>

$$x_{F_2} F_2 = x_{T_2} T_2 + x_{B_2} B_2 = 0.985 \times 7,300 + 0.01 \times 12,000 = 7,310.5 \quad (2)$$

## Toluene Balance

$$x_{F_2} F_2 = x_{T_2} T_2 + x_{B_2} B_2 = 0.015 \times 7,300 + 0.985 \times 12,000 = 11,929.5 \quad (3)$$

## Xylene Balance

$$x_{F_2} F_2 = x_{T_2} T_2 + x_{B_2} B_2 = 0 \times 7,300 + 0.005 \times 12,000 = 60 \quad (4)$$

Check balance: $F_2 = 7,310.5 + 11,929.5 + 60 = 19,300$ (Checks).

Since all streams have about the same liquid density, composition of stream $F_2$ is

$$Percent\ benzene = \frac{7310.5 \times 100}{19,300} = 37.88\% \quad (5)$$

$$Percent\ toluene = \frac{11,929.5 \times 100}{19,300} = 61.81\% \quad (6)$$

$$Percent\ xylene = \frac{60 \times 100}{19,300} = 0.31\% \quad (7)$$

These percentages add up to 100%.

## Part (a):  Material Balance around Envelope E

Because the holding tank level changed from 1.0 m to 1.5 m in a 24 hour period, $T_1$ is no longer equal to $F_2$ but is given by

$$T_1 = F_2 + \Delta I \quad (8)$$

where $\Delta I$ is the change in inventory. Thus the rate of flow from the top of tower 1 is

$$T_1 = F_2 + \Delta I = 19,300 + \frac{1\ gal}{mm} \times (1,500 - 1,000) mm = 19,800\ gallons \quad Ans.$$

## Part (b):

As in part (a) when the holding tank level varied from 1.0 m to 1.5 m during the 24 hour period, $T_1 = 19,800$ gallons. We need to know the composition of the $T_1$ stream to be able to calculate the composition of the feed to tower 1, $F_1$. Since the amounts of each component in stream $T_1$ must equal its amount in stream $F_2$, it follows that each component

214

composition must be obtained from

$$x_{T_1} T_1 = x_{F_2} F_2 + x_{F_2} \Delta I \quad or \quad x_{T_1} = \frac{x_{F_2}(F_2 + \Delta I)}{T_1} = x_{F_2} \quad (9)$$

Thus $T_1$ has the same composition as $F_2$. Since we were given that $B_1 = 10,000$ gallons, then, assuming similar densities for all streams

## Material Balance around Envelope C :

The volume of the feed, assuming similar densities and ideal solution, is
$$F_1 = T_1 + B_1 = 19,800 + 10,000 = 29,800 \ gallons \qquad Ans.$$
The composition of the feed stream, $F_1$, is obtained from

$$x_{F_1} F_1 = x_{T_1} T_1 + x_{B_1} B_1 \quad or \quad x_{F_1} = \frac{x_{T_1} T_1 + x_{B_1} B_1}{F_1} \ x100\% \qquad (10)$$

Thus, for Benzene
$$x_{F_1} = \frac{x_{T_1} T_1 + x_{B_1} B_1}{F_1} x100\% = \frac{0.3788 \, x \, 19,800 + 0 \, x \, 10,000}{29,800} x100 = 25.17\% \ Ans.$$

For Toluene
$$x_{F_1} = \frac{x_{T_1} T_1 + x_{B_1} B_1}{F_1} x100\% = \frac{0.6181 \, x \, 19,800 + 0.007 \, x \, 10,000}{29,800} x100 = 41.30\% \ Ans.$$

For Xylene
$$x_{F_1} = \frac{x_{T_1} T_1 + x_{B_1} B_1}{F_1} x100\% = \frac{0.0031 \, x \, 19,800 + 0.993 \, x \, 10,000}{29,800} x100 = 33.53\% \ Ans.$$

Check feed composition balance as 100 % = 25.17 % + 41.30 % + 33.53 %. (Checks).

These equations raise the possibility of dependent and independent equations. For example, equation of envelope D is not independent as it is a combination of equations for envelopes C and E. A useful formula for determining the number of independent equations is

*Number of equations = Number of components + Number of process units* (5.9)

## 5.2.1: Estimating Process Paramaters

## Illustrative Example 5.7: Vaporization Of Ideal Liquid Solutions

a): Estimate the partial pressure of an equi-molar benzene/toluene solution at 60 C given that for benzene, $P_A{}^0 = 388.6$ mm. Hg and for toluene, $P_B{}^0 = 139.5$ mm Hg.

b): What is the mole % of benzene in the vapour in equilibrium with the equi-molar solution?

c): If the vapours of part (b) were liquefied, what would be the compositions of the liquid and that of the vapour above it?

**Answer**

Part (a)

The partial pressures of benzene and toluene at 60 C are, from Raoult's law,

$$P_A = x_A \cdot P_A^o = 0.5 \times 388.6 = 194.3 \, mm \, Hg. \qquad\qquad Ans$$

$$P_B = x_B \cdot P_B^o = 0.5 \times 139.5 = 69.75 \, mm \, Hg. \qquad\qquad Ans$$

Part (b)

The total pressure in the system, $P_T$, is given by

$$P_T = P_A + P_B = 194.3 + 69.75 = 264.05 \, mm \, Hg. \qquad (1)$$

The number of moles of an ideal gas, n, is

$$n = \frac{PV}{RT} \qquad\qquad (2)$$

Thus the total number of moles in the system, $n_T$, is

$$n_T = n_A + n_B = \frac{P_A V}{RT} + \frac{P_B V}{RT} = (P_A + P_B) \cdot \frac{V}{RT} = \frac{P_T V}{RT} \qquad (3)$$

Thus the mole % of benzene and toluene are, respectyively,

$$y_A = \frac{n_A}{n_T} = \frac{P_A}{P_T} = \frac{194.3}{264.05} = 0.736 \qquad\qquad Ans$$

$$y_B = \frac{n_B}{n_T} = \frac{P_B}{P_T} = \frac{69.75}{264.05} = 0.264 \qquad\qquad Ans$$

Part (c )

When liquefied, the liquid mole fractions for benzene would be 0.736 and that for toluene, 0.264, since it is the same vapour that is now condensed into liquid. For the vapour above this liquid, however, Raoult's law gives the partial pressures as

$$P_A = x_A \cdot P_A^o = 0.736 \times 388.6 = 286.01 \, mm \, Hg. \qquad\qquad Ans$$

$$P_B = x_B \cdot P_B^o = 0.264 \times 139.5 = 36.83 \, mm \, Hg. \qquad\qquad Ans$$

216

The composition of the vapour is obtained from
$$P_T = P_A + P_B = 286.01 + 36.83 = 322.84 \, mm \, Hg. \tag{4}$$
and

$$y_A = \frac{n_A}{n_T} = \frac{P_A}{P_T} = \frac{286.01}{322.84} = 0.886 \qquad\qquad Ans$$

$$y_B = \frac{n_B}{n_T} = \frac{P_B}{P_T} = \frac{36.83}{322.84} = 0.114 \qquad\qquad Ans$$

Note that the mole fractions have increased.

If the above vapour is liquefied again, the partial pressures and mole fractions of the vapour above the now condensed liquid will be given as

$$P_A = x_A.P_A^o = 0.886 \, x \, 388.6 = 344.30 \, mm \, Hg. \qquad Ans$$

$$P_B = x_B.P_B^o = 0.114 \, x \, 139.5 = 15.90 \, mm \, Hg. \qquad Ans$$

The composition of the vapour is obtained from
$$P_T = P_A + P_B = 344.30 + 15.90 = 360.20 \, mm \, Hg. \tag{5}$$
and

$$y_A = \frac{n_A}{n_T} = \frac{P_A}{P_T} = \frac{344.30}{360.20} = 0.956 \qquad\qquad Ans$$

$$y_B = \frac{n_B}{n_T} = \frac{P_B}{P_T} = \frac{15.90}{360.2} = 0.044 \qquad\qquad Ans$$

This simple example illustrates the rationale for multi-stage fractional distillation or rectification, where the concentration of the lighter component in the vapour phase continues to increase with successive condensation and vaporisations.

### 5.2.2: Material Balance by the Method of Constraints

This is a general procedure which is able to handle material balance problems in which the mixtures may not be ideal solutions. In this method, a large number of equations is reduced, in most cases, to a few which are, then, solved by any number of analytical or numerical solution procedures. This method is described in detail in the book by Myers & Seider (1976) and only the basic procedure is outlined here. Four stages are involved.

Stage 1: Determine the equations and constraints of the process

The equations are (a) the material balance equations, (b) the mole fraction equations and (c) the equipment constraints or parameters. Equipment constraints are process variables, other than stream parameters, such as flowrate or composition, which constrain the operation of a particular unit. A typical equipment constraint could be the ratio of inflow to outflow or an equilibrium constant.

Stage 2: <u>Determine the number of variables involved</u>

If there are $N_S$ streams, $N_C$ mole fractions and $N_P$ equipment parameters, the number of variables, $N_V$, is given by

$$N_V = N_S(N_C + 1) + N_P \qquad (5.10)$$

Stage 3: <u>Determine the design determining (decision) variables</u>

The decision variables are, usually, those variables whose values are, either, most easily, determined, most frequently used or those which affect process economics or operation directly or significantly. If there are $N_E$ equations, the number of decision variables, $N_D$, is given by

$$N_D = N_V - N_E \qquad (5.11)$$

Stage 4: <u>Obtain the solution</u>

The solution is obtained by reducing the material balance equations which, often, turn out to be linear or non -linear simultaneous equations, to a few equations with one or two unknowns. This is done using a combination of the decision variables, mole fractions and equipment constraints. The resulting equations are, then, solved using any one of (a) precedence ordering methods (b) matrix methods (c) trial and error methods.

To illustrate the method, consider a binary system of components 1 and 2 undergoing distillation as shown in the sketch below.

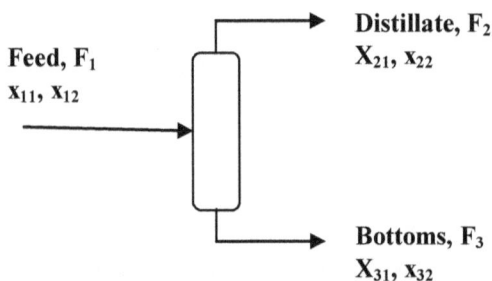

The streams are coded by $F_i$ and composition by $x_{ij}$ where i represents the stream and j the component. For example, if feed stream is coded i = 1, distillate stream is coded i = 2 and the bottoms stream is coded i = 3 while component 1 is coded j = 1 and component 2, j = 2, the four stages of the method are seen to yield the following:

### 1. Equations and Constraints

### (a) Material Balance Equations

| | | |
|---|---|---|
| Overall balance | $F_1 = F_2 + F_3$ | (5.12) |
| Component 1 balance: | $F_1 x_{11} = F_2 x_{21} + F_3 x_{31}$ | (5.13) |
| Component 2 balance: | $F_1 x_{12} = F_2 x_{22} + F_3 x_{32}$ | (5.14) |

### (b) Mole Fraction Constraints

| | | |
|---|---|---|
| Feed stream: | $x_{11} + x_{12} = 1$ | (5.15) |
| Distillate stream: | $x_{21} + x_{22} = 1$ | (5.16) |
| Bottoms stream: | $x_{31} + x_{32} = 1$ | (5.17) |

### (c) Equipment Constraints

In this particular example, none is specified.

### 2. Number of Variables

From equation (5.10)
$$N_V = N_S(N_C + 1) + N_P = 3(2 + 1) + 0 = 9 \tag{5.18}$$
These are $F_1$, $F_2$, $F_3$, $x_{11}$, $x_{12}$, $x_{21}$, $x_{22}$, $x_{31}$, $x_{32}$

### 3. Design/Decision Variables

From equation (5.11)
$$N_D = N_V - N_E = 9 - 5 = 4 \tag{5.19}$$
These can be chosen as $F_1$, $x_{11}$, $x_{21}$ and $x_{31}$

### 4. Solution

The order in which the decision/deasign variables are solved for depends on which of the other variables are known or given.

**5.2.2.1: Types of Equipment Constraints**

219

Three major types of equipment constraints are, usually encountered.

a). <u>Constraints not imposed by purely physical processes (inventory, assembly or separation operations)</u>

These are constraints based on human choice and not because of some restriction imposed by natural processes. Such a constraint may, however, be of an economic nature and may be expressed, for example, as a stream ratio, say $\alpha = F_1/F_2$.

b). <u>Constraints imposed by purely natural processes (inventory, assembly or sepation operations)</u>

This type of constraint is usually based on some equilibrium relationship or value of a physical constant beyond which a variable or a particular set of variables cease to behave in an accustomed manner.
One of the most important, and typical, of this type of constraint is the equilibrium constant which has several variations of the general definition depending on the characteristics of the physical system.

The general definition of the equilibrium constant in a physical system, with respect to the jth component, is

$$K_j = \frac{x_{Lj}}{x_{Hj}} \qquad (5.20)$$

where $x_{Lj}$ is the composition of the component j in the lighter phase and $x_{Hj}$ is its composition in the heavier phase.

This definition becomes modified for particular systems as follows

i)      For systems in which insoluble gases are involved such as hydrogen in water, $x_{Hj} = 0$ and
$$K_j = \infty \qquad (5.21)$$

ii)     For systems which involve dissolved solids or non-volatile liquids such as salt in water or non-volatile oil in water,
$$x_{Lj} = 0 \text{ and } K_j = 0 \qquad (5.22)$$

iii)    For systems in which ideal solutions, such as a benzene/toluene mixture, to which Raoult's law applies, are involved,

$$K_j = \frac{P_j^o}{P_T} \qquad (5.23)$$

where $P_j^o$ is the saturation vapour pressure of component j
and $P_T$ is the total system pressure.

iv)      For systems of non-ideal solutions to which Henry's law
applies

$$K_j = \frac{H_j}{P_T} \qquad (5.24)$$

where $H_j$ is the Henry's law constant for componenmt j and
$P_T$ is the total system pressure.

## c). Constraints imposed by chemical reactions

In this case, the constraint arises from the limits set by chemical
reactions with the equilibrium constant defined as

$$K_P = P^{\sum_j \nu_j} \prod x_j^{\nu_j} \qquad (5.25)$$

where P is the total system pressure, $\sum_j \nu_j$ the algebraic sum of the

stoichiometric coefficients of reactants and products, $\nu_j$ the
stoichiometric coefficient of component j. For example, the reaction

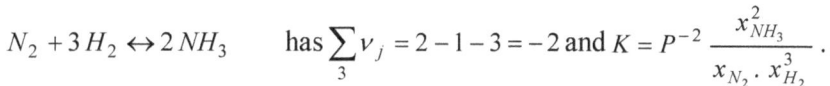

$$N_2 + 3H_2 \leftrightarrow 2NH_3 \qquad \text{has} \sum_3 \nu_j = 2 - 1 - 3 = -2 \text{ and } K = P^{-2} \frac{x_{NH_3}^2}{x_{N_2} \cdot x_{H_2}^3}.$$

Note that when K < 1, the reaction, from left to right as written, is not
feasible. When K = 1, the reaction is reversible whilst if K > 1, the
reaction is irreversible.

## Illustrative Example 5.8

## No constraint imposed by the physical system

A mixture of benzene ($C_6H_6$) and carbon tetrachloride ($CCl_4$) is to be
separated by distillation. Carbon tetrachloride is slightly more volatile
than benzene. During continuous operation of the distillation tower, the
flow rates and compositions shown in the Table below were observed.
Calculate the flowrate and composition of the bottom stream.

| Stream | Flow rate, kmol/h | Mole fraction CCl$_4$ |
|---|---|---|
| Feed | 5.26 | 0.445 |
| Distillate | 2.32 | 0.932 |

## Answer

The system may be represented as shown below as a binary system of components 1 (CCl$_4$) and 2 (benzene) undergoing distillation. The feed stream is coded $F_1$, distillate stream is coded $F_2$ and the bottoms stream is coded $F_3$. Thus $F_1$ = 5.26 kmol/h, $F_2$ = 2.32 kmol/h, $x_{11} = 0.445$, $x_{21} = 0.932$.

From equation (5.12), $F_3 = F_1 - F_2 = 5.26 - 2.32 = 2.94 \, kmol \, / \, h$.

From equation (5.13)

$$x_{31} = \frac{F_1 \, x_{11} - F_2 \, x_{21}}{F_3} = \frac{5.26 \, x \, 0.445 - 2.32 \, x \, 0.932}{2.94} = 0.061$$

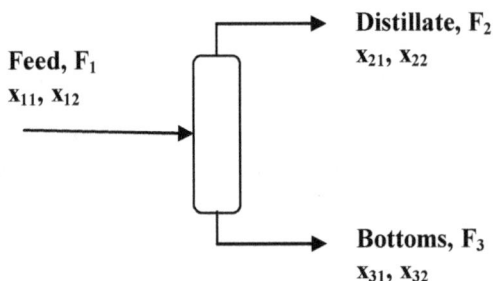

From equation (5.17)

$$x_{32} = 1 - x_{31} = 1 - 0.061 = 0.939$$

Hence the flow rate and composition of the bottom stream is

| Flow rate, kmol/h | 2.94 | |
|---|---|---|
| Mole fraction CCl$_4$ | 0.061 | |
| Mole fraction benzene | 0.939 | Ans |

## 5.2.2.1.1: Physical Processes restricted by some Natural Phenomenon

Let us consider the case in which the restriction arises from the equilibrium constant. For ease of analysis and clarity, let us consider a compound which is to be separated from a feed consisting of it and three other different compounds. All of them are present in both the overhead and bottom streams but such that the overhead stream is more concentrated in the desired component than the bottoms stream which is more concentrated in the heavier components. Let us develop a solution for the material balance for this process.

**Answer**

The diagram for the process, including the nomenclature, is illustrated below.

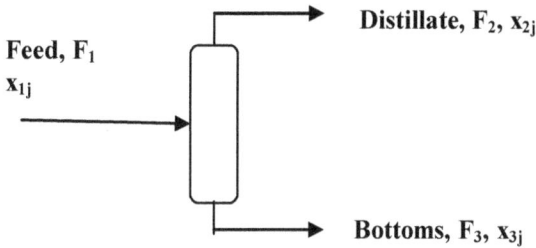

1    Equations and Constraints

a)  Material Balance Equations

$$F_1 = F_2 + F_3 \tag{5.26}$$

Overall balance

Component 1 balance:  $F_1 x_{11} = F_2 x_{21} + F_3 x_{31}$ (5.27)

Component 2 balance:  $F_1 x_{12} = F_2 x_{22} + F_3 x_{32}$ (5.28)

Component 3 balance:  $F_1 x_{13} = F_2 x_{23} + F_3 x_{33}$ (5.29)

Component 4 balance:  $F_1 x_{14} = F_2 x_{24} + F_3 x_{34}$ (5.30)

(b)  Mole Fraction Constraints

Feed stream:    $x_{11} + x_{12} + x_{13} + x_{14} = 1$ (5.31)

Distillate stream:  $x_{21} + x_{22} + x_{23} + x_{24} = 1$ (5.32)

Bottoms stream:    $x_{31} + x_{32} + x_{33} + x_{34} = 1$ (5.33)

(c )  Equipment Constraints

$$x_{21} = k_1 x_{31} \tag{5.34}$$

$$x_{22} = k_2 x_{32} \tag{5.35}$$

$$x_{23} = k_3 x_{33} \tag{5.36}$$

223

$$x_{24} = k_4 x_{34} \qquad (5.37)$$

## 2. Number of Variables

From equation (5.10)
$$N_V = N_S(N_C + 1) + N_P = 3(4 + 1) + 4 = 19 \qquad (5.38)$$

## 3. Design/Decision Variables

From equation (5.11)
$$N_D = N_V - N_E = 19 - 11 = 8 \qquad (5.39)$$
These can be chosen as $F_1$, $x_{11}$, $x_{12}$, $x_{14}$, $k_1$, $k_2$, $k_3$, $k_4$.

## 4. Solution

From the statement of the problem, $x_{11}$, $x_{12}$ and $x_{14}$ are known so that, from equation (5.33)
$$x_{13} = 1 - x_{11} - x_{12} - x_{14} \qquad (5.40)$$
Equations (5.27) to (5.30) may, also, be expressed, generally as, for j = 1 to 4,
$$F_1 x_{1j} = F_2 x_{2j} + F_3 x_{3j} \qquad (5.41)$$
Similarly, equations (5.34) to (5.37) can, also, be expressed generally, for j = 1 to 4, as
$$x_{2j} = k_j x_{3j} \qquad (5.42)$$
Substituting for $x_{2j}$ from equations (5.42) into equation (5.41) and solving for $x_{3j}$
$$F_1 x_{1j} = F_2 k_j x_{3j} + F_3 x_{3j} = x_{3j}\left(F_2 k_j + F_3\right)$$
That is
$$x_{3j} = \frac{F_1 x_{1j}}{\left(F_2 k_j + F_3\right)} = \frac{(F_1/F_2)\, x_{1j}}{\left(k_j + F_3/F_2\right)} \qquad (5.43)$$
Thus
$$\sum_{j=1}^{4} x_{3j} = 1 = \sum_{j=1}^{4} \frac{(F_1/F_2)\, x_{1j}}{\left(k_j + F_3/F_2\right)} \qquad (5.44)$$

From (5.26), putting $F_3/F_2 = \alpha$
$$\frac{F_1}{F_2} = 1 + \frac{F_3}{F_2} = 1 + \alpha \qquad (5.45)$$
Equation (5.44)), thus, becomes

224

$$\sum_{j=1}^{4} x_{3j} = 1 = \sum_{j=1}^{4} \frac{(1+\alpha) x_{1j}}{(k_j + \alpha)} \tag{5.46}$$

We can, also, substitute for $x_{3j}$ from equations (5.42) into equation (5.41) and solve for $x_{2j}$

$$F_1 x_{1j} = F_2 x_{2j} + F_3 \frac{x_{2j}}{k_j} = x_{2j} \left( \frac{F_2 k_j + F_3}{k_j} \right)$$

That is

$$x_{2j} = \frac{F_1 k_j x_{1j}}{(F_2 k_j + F_3)} = \frac{(F_1/F_2) k_j x_{1j}}{(k_j + F_3/F_2)} \tag{5.47}$$

Thus

$$\sum_{j=1}^{4} x_{2j} = 1 = \sum_{j=1}^{4} \frac{(F_1/F_2) k_j x_{1j}}{(k_j + F_3/F_2)} \tag{5.48}$$

Subtracting equation (5.48) from equation (5.44)

$$0 = \sum_{j=1}^{4} \frac{(F_1/F_2) x_{1j}}{(k_j + F_3/F_2)} - \sum_{j=1}^{4} \frac{(F_1/F_2) k_j x_{1j}}{(k_j + F_3/F_2)} = \sum_{j=1}^{4} \frac{(F_1/F_2) x_{1j} (1 - k_j)}{k_j + F_3/F_2} \tag{5.49}$$

Since $F_3/F_2 = \alpha$, then

$$\sum_{j=1}^{4} \frac{x_{1j} (1 - k_j)}{k_j + \alpha} = 0 \tag{5.50}$$

in which $\alpha$ is the only unknown. Equation (5.46) or (5.50) may be solved by either a graphical procedure or Newton's iteration method such that if, from (5.46)

$$f(\alpha) = 1 - \sum_{j=1}^{4} \frac{(1+\alpha) x_{1j}}{(k_j + \alpha)} \tag{5.51}$$

Or from (5.50)

$$f(\alpha) = \sum_{j=1}^{4} \frac{x_{1j} (1 - k_j)}{k_j + \alpha} \tag{5.52}$$

and the starting value of $\alpha$ is designated as $\alpha_{old}$, the new $\alpha$ designated as $\alpha_{new}$ is given by

$$\alpha_{new} = \alpha_{old} - \frac{f(\alpha_{old})}{f'(\alpha_{old})} \tag{5.53}$$

$f'(\alpha)$ is the first derivative of $f(\alpha)$ with respect to $\alpha$ given by, from (5.46)

$$f'(\alpha) = \sum_{j=1}^{4} \frac{x_{1j} (1 - k_j)}{(k_j + \alpha)^2} \tag{5.54}$$

225

and from (5.50)

$$f'(\alpha) = -\sum_{j=1}^{4} \frac{x_{1j}\left(1-k_j\right)}{\left(k_j+\alpha\right)^2} \tag{5.55}$$

Convergence is achieved when some acceptable level of error, $\varepsilon$, defined below, is obtained.

$$\varepsilon \geq \left|\frac{\alpha_{new}-\alpha_{old}}{\alpha_{old}}\right| \tag{5.56}$$

**Illustrative Example 5.9**

10 kmol/h of a feed, consisting of four components whose feed concentrations are known, is to be separated in a tower into an overhead and a bottoms stream. If the feed composition and equilibrium constants of the substances, designated 1 to 4, are as given below, calculate the flow rates and compositions of the overhead and bottom streams for an error, $\varepsilon$, not more than 0.005.

| Component | 1 | 2 | 3 | 4 |
|---|---|---|---|---|
| Feed composition | 0.10 | 0.80 | 0.06 | 0.04 |
| Equilibrium constant | 2 | 0.8 | 1.4 | 2.5 |

**Answer**

From (5.52)

$$f(\alpha) = \sum_{j=1}^{4} \frac{x_{1j}\left(1-k_j\right)}{k_j+\alpha} \tag{1}$$

From (5.55)

$$f'(\alpha) = -\sum_{j=1}^{4} \frac{x_{1j}\left(1-k_j\right)}{\left(k_j+\alpha\right)^2} \tag{2}$$

From (5.53)

$$\alpha_{new} = \alpha_{old} - \frac{f\left(\alpha_{old}\right)}{f'\left(\alpha_{old}\right)} \tag{3}$$

If we guess an initial value of $\alpha$ to be 2, then from (1)

$$f(\alpha) = \frac{0.10\left(1-2\right)}{2+2} + \frac{0.80\left(1-0.8\right)}{0.8+2} + \frac{0.06\left(1-1.4\right)}{1.4+2} + \frac{0.04\left(1-2.5\right)}{2.5+2}$$

$$= 0.0118 \tag{4}$$

From (2)

$$f'(\alpha) = -\frac{0.10\,(1-2)}{(2+2)^2} - \frac{0.80\,(1-0.8)}{(0.8+2)^2} - \frac{0.06\,(1-1.4)}{(1.4+2)^2} - \frac{0.04\,(1-2.5)}{(2.5+2)^2}$$

$$= -0.0091 \tag{5}$$

From (3)

$$\alpha_{new} = 2 - \frac{0.0118}{-0.0091} = 3.297 \tag{6}$$

From (5.56)

$$\varepsilon = \left| \frac{3.297 - 2}{2} \right| = 0.648 \gg 0.005 \tag{7}$$

This is obviously unacceptable. We use a new value of $\alpha = 3.297$, then from (1)

$$f(\alpha) = \frac{0.10\,(1-2)}{2+3.297} + \frac{0.80\,(1-0.8)}{0.8+3.297} + \frac{0.06\,(1-1.4)}{1.4+3.297} + \frac{0.04\,(1-2.5)}{2.5+3.297}$$

$$= 0.0047 \tag{8}$$

From (2)

$$f'(\alpha) = -\frac{0.10\,(1-2)}{(2+3.297)^2} - \frac{0.80\,(1-0.8)}{(0.8+3.297)^2} - \frac{0.06\,(1-1.4)}{(1.4+3.297)^2} - \frac{0.04\,(1-2.5)}{(2.5+3.297)^2}$$

$$= 0.0160 \tag{9}$$

From (3)

$$\alpha_{new} = 3.297 - \frac{0.0047}{0.0160} = 3.003 \tag{10}$$

From (5.56)

$$\varepsilon = \left| \frac{3.003 - 3.297}{3.297} \right| = 0.0891 \gg 0.005 \tag{11}$$

This is still unacceptable. We use a new value of $\alpha = 3$, then from (1)

$$f(\alpha) = \frac{0.10\,(1-2)}{2+3} + \frac{0.80\,(1-0.8)}{0.8+3} + \frac{0.06\,(1-1.4)}{1.4+3} + \frac{0.04\,(1-2.5)}{2.5+3}$$

$$= 0.0057 \tag{12}$$

From (2)

$$f'(\alpha) = -\frac{0.10\,(1-2)}{(2+3)^2} - \frac{0.80\,(1-0.8)}{(0.8+3)^2} - \frac{0.06\,(1-1.4)}{(1.4+3)^2} - \frac{0.04\,(1-2.5)}{(2.5+3)^2}$$

$$= -0.0039 \tag{13}$$

From (3)

$$\alpha_{new} = 3 - \frac{0.0057}{-0.0039} = 4.462 \tag{14}$$

From (5.56)

$$\varepsilon = \left| \frac{4.462 - 3}{3} \right| = 0.487 \gg 0.005 \tag{15}$$

We use a new value of $\alpha = 4.462$, then from (1)

$$f(\alpha) = \frac{0.10\,(1-2)}{2+4.462} + \frac{0.80\,(1-0.8)}{0.8+4.462} + \frac{0.06\,(1-1.4)}{1.4+4.462} + \frac{0.04\,(1-2.5)}{2.5+4.462}$$

$$= 0.0022 \tag{16}$$

From (2)

$$f'(\alpha) = -\frac{0.10\,(1-2)}{(2+4.462)^2} - \frac{0.80\,(1-0.8)}{(0.8+4.462)^2} - \frac{0.06\,(1-1.4)}{(1.4+4.462)^2} - \frac{0.04\,(1-2.5)}{(2.5+4.462)^2}$$

$$= -0.0015 \tag{17}$$

From (3)

$$\alpha_{new} = 4.462 - \frac{0.0022}{-0.0015} = 5.929 \tag{18}$$

From (5.56)

$$\varepsilon = \left| \frac{5.929 - 4.462}{4.462} \right| = 0.3288 \gg 0.005 \tag{19}$$

We use a new value of $\alpha = 5.929$, then from (1)

$$f(\alpha) = \frac{0.10\,(1-2)}{2+5.929} + \frac{0.80\,(1-0.8)}{0.8+5.929} + \frac{0.06\,(1-1.4)}{1.4+5.929} + \frac{0.04\,(1-2.5)}{2.5+5.929}$$

$$= 0.0008 \tag{20}$$

From (2)

$$f'(\alpha) = -\frac{0.10\,(1-2)}{(2+5.929)^2} - \frac{0.80\,(1-0.8)}{(0.8+5.929)^2} - \frac{0.06\,(1-1.4)}{(1.4+5.929)^2} - \frac{0.04\,(1-2.5)}{(2.5+5.929)^2}$$

$$= -0.0007 \tag{21}$$

From (3)

$$\alpha_{new} = 5.929 - \frac{0.0008}{-0.0007} = 7.072 \tag{22}$$

From (5.56)

$$\varepsilon = \left| \frac{7.072 - 5.929}{5.929} \right| = 0.1928 \gg 0.005 \tag{23}$$

Because the Newton's method will not converge, as can be seen from the graphical plot of the data, a graphical solution of equation (5.52) and (5.51) is used. The correct answer is obtained when $f(\alpha) = 0$. This occurs, in both cases, at $\alpha = 5.7$.

**Graphical Solution of Equation 5.52 for Illustrative Example 5.9**

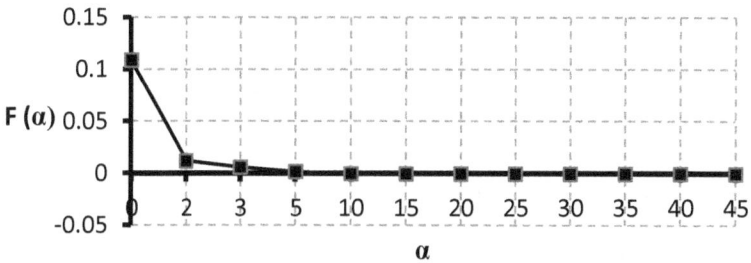

**Graphical Solution of Equation 5.51 for Illustrative Example 5.9**

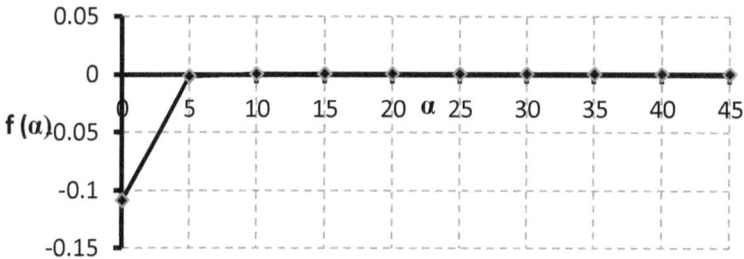

This is confirmed when f($\alpha$) is recalculated using $\alpha = 5.7$, to be

$$f(\alpha) = \frac{0.10\,(1-2)}{2+5.7} + \frac{0.80\,(1-0.8)}{0.8+5.7} + \frac{0.06\,(1-1.4)}{1.4+5.7} + \frac{0.04\,(1-2.5)}{2.5+5.7} = 0.0009 \quad (24)$$

From (5,45)

$$F_2 = \frac{F_1}{1+\alpha} = \frac{10}{6.7} = 1.493\,kmol\,/\,h \quad Ans \quad (25)$$

$$F_3 = \alpha\,F_2 = 5.7\,x\,1.493 = 8.510\,kmol\,/\,h \quad Ans \quad (26)$$

The compositions of the bottomsn and distillate streams are obtained from the component balance and equilibrium equations (5.27) and (5.34) as

$$F_1\,x_{11} = F_2\,k_1 x_{31} + F_3\,x_{31} = x_{31}\left(F_2\,k_1 + F_3\right)$$

$$x_{31} = \frac{F_1\,x_{11}}{\left(F_2\,k_1 + F_3\right)} = \frac{10\,x\,0.1}{1.493\,x\,2 + 8.510} = 0.087 \quad (27)$$

Similarly

229

$$x_{32} = \frac{F_1 x_{12}}{(F_2 k_2 + F_3)} = \frac{10 \times 0.8}{1.493 \times 0.8 + 8.510} = 0.824 \qquad (28)$$

$$x_{33} = \frac{F_1 x_{13}}{(F_2 k_3 + F_3)} = \frac{10 \times 0.06}{1.493 \times 1.4 + 8.510} = 0.057 \qquad (29)$$

$$x_{34} = \frac{F_1 x_{14}}{(F_2 k_4 + F_3)} = \frac{10 \times 0.04}{1.493 \times 2.5 + 8.510} = 0.033 \qquad (30)$$

So that for the bottoms stream

$$x_{31} + x_{32} + x_{33} + x_{34} = 0.087 + 0.824 + 0.057 + 0.033 = 1.001 \quad checks \quad (31)$$

From the equilibrium constraints

$$x_{21} = k_1 x_{31} = 2 \times 0.087 = 0.174 \qquad (32)$$

$$x_{22} = k_2 x_{32} = 0.8 \times 0.824 = 0.659 \qquad (33)$$

$$x_{23} = k_3 x_{33} = 1.4 \times 0.057 = 0.080 \qquad (34)$$

$$x_{24} = k_4 x_{34} = 2.5 \times 0.033 = 0.083 \qquad (35)$$

For the distillate stream

$$x_{21} + x_{22} + x_{23} + x_{24} = 0.174 + 0.659 + 0.080 + 0.083 = 0.996 \quad checks \quad (36)$$

### 5.2.2.1.2: *Processes restricted by Chemical Reaction*

When chemical reactions occur, the procedure is, essentially, the same except that the material balance equations are developed differently in order to account for the consumption and production of species which result from the chemical reaction. The treatment here is that due to Myers & Seider, (1976).

The first step is to write out a balanced chemical reaction equation such as the one set below.

$$v_A A + v_B B \leftrightarrow v_C C + v_D D \qquad (5.57)$$

for which the material balance is

$$\sum_{j=1}^{N} v_j C_j = 0 \quad for\ compounds \qquad (5.58)$$

$$\sum_{j=1}^{N} v_j m_{jk} = 0 \quad for\ elements \qquad (5.59)$$

where $C_j$ is the jth compound .in a set of N compounds, $m_{jk}$ is the number of atoms of the kth element.in the jth compound and $v_j$ is the stoichiometric coefficient of the jth compound.

Secondly, the rate of chemical reaction is related to the rate of production or consumption of any particular species by the stoichiometric coefficient of that species. Thus, if r is the rate of a chemical reaction in which $n_j$ moles of species j, having a stoichiometric coefficient of $v_j$ is involved, then

$$\frac{d n_j}{d t} = r v_j \qquad (5.60)$$

where $\dfrac{d n_j}{d t}$ is the rate of production or collsumption of species, j.

**Illustrative Example 5.10**

In the simple chemical reaction,

$$H_2 + \tfrac{1}{2} H_2 O = H_2 O$$

calculate the rate of consumption of $H_2$ and $O_2$ if the rate of production of $H_2O$ is 5 mols/h.

**Answer**

From equation (5.60) and the data given

$$r = \frac{1}{v_j} \frac{d n_j}{d t} = \frac{1}{1} x\, 5\, mols\,/\,h = 5\, mols\,/\,h \qquad Ans$$

For $H_2$

$$\frac{d n_j}{d t} = r v_j = 5\, mols\,/\,h\, x\, 1 = 5\, mols\,/\,h\, (consumption) \quad Ans$$

For $O_2$

$$\frac{d n_j}{d t} = r v_j = 5\, mols\,/\,h\, x\, \frac{1}{2} = 2.5\, mols\,/\,h\, (consumption) \quad Ans$$

Thirdly, the general material balance for an open system is given by

**INPUT + GENERATION = OUTPUT + ACCUMULATION**

At steady state, accumulation is zero, so that if $F_{IN}$ represents input flow of the species into the process, $F_{GEN}$ the flow generated .in the process and $F_{OUT}$ the flow out of the process, then

$$F_{IN} + F_{GEN} = F_{OUT} \qquad (5.61)$$

From (5.60)

$$F_{GEN} = \frac{d n_j}{d t} = r v_j \qquad (5.62)$$

An ordinary material balance of all species in the system gives

$$F_{IN} - F_{OUT} = \sum F_i \, x_{ij} \qquad (5.63)$$

Combining equations (5.61), (5.62) and (5.63), for each j, $N_S$ species,

$$\sum_{i=1}^{N_S} F_i \, x_{ij} + r v_j \qquad (5.64)$$

A similar analysis for the elements gives

$$\sum_{j=1}^{N_C} \sum_{i=1}^{N_S} F_i \, x_{ij} m_{jk} + r \sum_{j=1}^{N_C} v_j m_{jk} = 0 \qquad (5.65)$$

But from (5.59)

$$\sum_{j=1}^{N} v_j m_{jk} = 0 \quad \text{for elements} \qquad \qquad \text{from} \qquad (5.59)$$

Hence

$$\sum_{j=1}^{N_C} \sum_{i=1}^{N_S} F_i \, x_{ij} m_{jk} = 0 \qquad (5.66)$$

Equation (5.66) is the equation to be solved for the mass balance and is called the stoichiometric equation (Myers & Seider, 1976). The following example from Myers & Seider, (1976), illustrates the use of equation (5.66).

**Illustrative Example 5.11**

Derive the expressions for the material balance in a burner which burns a fuel in air. The general formula for the fuel is given by $C_k H_r O_p N_q S_m$ where C is for carbon, H for hydrogen, N for nitrogen and S is for sulphur and k, p, q, r, and m are constants.

**Answer**           .

The burner may be represented schematically as follows:

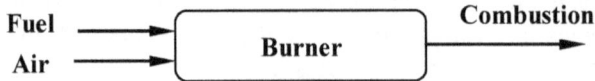

The combustion reaction equation may be represented as

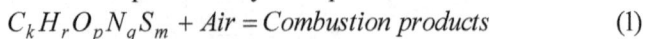

$$C_k H_r O_p N_q S_m + Air = Combustion \ products \qquad (1)$$

Since trying to balance the equation with unknown coefficients is not practical. Instead, let us code the streams and components as shown in the Table below.

| Stream | Fuel | Air | Products | | | |
|---|---|---|---|---|---|---|
| Code | 1 | 2 | 3 | | | |
| Component | Fuel | $O_2$ | $CO_2$ | $H_2O$ | $N_2$ | $SO_2$ |
| Code | 1 | 2 | 3 | 4 | 5 | 6 |

## 1.  Material balance equations

If we assume steady state and complete combustion, then from (5.66), the material balance on each species is as follows

Fuel  $\qquad F_3\, x_{31} = 0$  (2)

Hydrogen  $r\, F_1 = 2\, F_3\, x_{34}$  (3)

Carbon  $k\, F_1 = F_3\, x_{33}$  (4)

Oxygen  $p\, F_1 + 2\, F_2\, x_{22} = 2\, F_3 x_{32} + 2\, F_3\, x_{33} + F_3 x_{34} + 2\, F_3\, x_{36}$  (5)

Nitrogen  $q\, F_1 + 2\, F_2\, x_{25} = 2\, F_3 x_{35}$  (6)

Sulphur  $m\, F_1 = F_3\, x_{36}$  (7)

## 2.  Mole fraction constraints

Fuel Stream  $\qquad x_{11} = 1$  (8)

Air Stream  $\qquad x_{22} + x_{25} = 1$  (9)

Products Stream  $\qquad x_{31} + x_{32} + x_{33} + x_{34} + x_{35} + x_{36} = 1$  (10)

## 3.  Equipment constraints

The equipment constraints are the air/fuel ratio, $\alpha$, and the equilibrium constant, where

$$\alpha = \frac{F_2}{F_1}$$  (11)

$$K = \infty$$  (12)

## 4.  Number of variables

Since there are 3 streams, 7 mole fractions and 1 constant (we cannot use $K = \infty$)

$$N_V = 11 \tag{13}$$

## 5.  Design/decision variables

Since there is complete combustion, $x_{31} = 0$ and the number of equations, from equations (3) to (10), is 8. Thus

$$N_D = N_V - N_E = 11 - 8 = 3 \tag{14}$$

## 6.  Solution of the problem

Since the number of moles, $n_{ij} = F_i x_{ij}$, the solution can be made easier, using this notation which helps to linearise the equations. Thus

equation 3 becomes   $r F_1 = 2 n_{34}$   $\hspace{2cm}$ (15)

equation 4 becomes   $k F_1 = n_{33}$   $\hspace{2cm}$ (16)

equation 5 becomes   $p F_1 + 2 n_{22} = 2 n_{32} + 2 n_{33} + n_{34} + 2 n_{36}$   $\hspace{1cm}$ (17)

equation 6 becomes   $q F_1 + 2 n_{25} = 2 n_{35}$   $\hspace{2cm}$ (18)

equation 7 becomes   $m F_1 = n_{36}$   $\hspace{2cm}$ (19)

To solve the linearised equations, multiply equation (10) by $F_3$ to get

$$F_3 x_{32} + F_3 x_{33} + F_3 x_{34} + F_3 x_{35} + F_3 x_{36} = F_3$$

or

$$F_3 = n_{32} + n_{33} + n_{34} + n_{35} + n_{36} \tag{20}$$

We can substitute the various values of the $n_{ij}$ of equations (15) to (19) into equation (20) to get

$$F_3 = \frac{p}{2} F_1 + n_{22} - k F_1 - \frac{r}{4} F_1 - m F_1 + k F_1 + \frac{r}{2} F_1 + \frac{q}{2} F_1 + n_{25} + m F_1$$

$$F_3 = \frac{p}{2} F_1 + n_{22} + \frac{r}{4} F_1 + \frac{q}{2} F_1 + n_{25} \tag{21}$$

Multiplying equation (9) by $F_2$ and using equation (11)

$$F_2 x_{22} + F_2 x_{25} = F_2 = n_{22} + n_{25} = \alpha F_1 \tag{22}$$

Substituting (22) into (21

$$F_3 = \frac{p}{2} F_1 + \frac{r}{4} F_1 + \frac{q}{2} F_1 + |\alpha F_1 = \left( \frac{2p + r + 2q + 4\alpha}{4} \right) F_1 \tag{23}$$

We can go back to equations (3) to (6) using equation (23) to get from (3) and (23)

$$x_{34} = \frac{r F_1}{2 F_3} = \frac{4 r F_1}{2(2 p + r + 2 q + 4 \alpha) F_1} = \frac{2 r}{(2 p + r + 2 q + 4 \alpha)} \quad (24)$$

From ((4) and (23)

$$x_{33} = \frac{k F_1}{F_3} = \frac{4 k F_1}{(2 p + r + 2 q + 4 \alpha) F_1} = \frac{4 k}{(2 p + r + 2 q + 4 \alpha)} \quad (25)$$

From ((7) and (23)

$$x_{36} = \frac{m F_1}{F_3} = \frac{4 m F_1}{(2 p + r + 2 q + 4 \alpha) F_1} = \frac{4 m}{(2 p + r + 2 q + 4 \alpha)} \quad (26)$$

From ((6) and (23)

$$x_{35} = \frac{q F_1 + 2 F_2 x_{25}}{2 F_3} = \frac{4(q F_1 + 2 \alpha F_1 x_{25})}{2(2 p + r + 2 q + 4 \alpha) F_1} = \frac{2(q + 2 \alpha x_{25})}{(2 p + r + 2 q + 4 \alpha)} \quad (27)$$

From (5)

$$p F_1 + 2 F_2 x_{22} = 2 F_3 x_{32} + 2 F_3 x_{33} + F_3 x_{34} + 2 F_3 x_{36}$$

That is

$$x_{32} = \frac{\left(p F_1 + 2 \alpha F_1 x_{22} - 2 F_3 x_{33} - F_3 x_{34} - 2 F_3 x_{36}\right)}{2 F_3}$$

$$= \frac{\left(p + 2 \alpha x_{22}\right) F_1}{2 F_3} - x_{33} - \frac{1}{2} x_{34} - x_{36}$$

$$= \frac{\left(p + 2 \alpha x_{22}\right) F_1}{2 F_3} - \frac{4 k}{(2 p + r + 2 q + 4 \alpha)} - \frac{1}{2} \frac{2 r}{(2 p + r + 2 q + 4 \alpha)}$$

$$- \frac{4 m}{(2 p + r + 2 q + 4 \alpha)}$$

$$= \frac{\left(p + 2 \alpha x_{22}\right) F_1}{2 F_3} - \frac{8 k + 2 r + 8 m}{2(2 p + r + 2 q + 4 \alpha)}$$

$$= \frac{4\left(p + 2 \alpha x_{22}\right) F_1}{2(2 p + r + 2 q + 4 \alpha)} - \frac{8 k + 2 r + 8 m}{2(2 p + r + 2 q + 4 \alpha)}$$

$$= \frac{2 p - 4 k - r - 4 m + 4 \alpha x_{22}}{(2 p + r + 2 q + 4 \alpha)} \quad (28)$$

Note that $x_{22}$ and $x_{25}$ are not unknowns as they are the mole fractions of oxygen and nitrogen in air, respectively.

As a check, since from (10)

$$x_{32} + x_{33} + x_{34} + x_{35} + x_{36} = 1$$

$$\frac{2p - 4k - r - 4m + 4\alpha x_{22}}{(2p + r + 2q + 4\alpha)} + \frac{4k}{(2p + r + 2q + 4\alpha)} + \frac{2r}{(2p + r + 2q + 4\alpha)}$$

$$+ \frac{2(q + 2\alpha x_{25})}{(2p + r + 2q + 4\alpha)} + \frac{4m}{(2p + r + 2q + 4\alpha)} = 1$$

$$= \frac{2p - 4k - r - 4m + 4\alpha x_{22} + 4k + 2r + 2q + 4\alpha x_{25} + 4m}{(2p + r + 2q + 4\alpha)}$$

$$= \frac{2p + r + 2q + 4\alpha(x_{22} + x_{25})}{(2p + r + 2q + 4\alpha)} = 1 \quad (checks) \tag{29}$$

## Illustrative Example 5.12

The composition of liquified petroleum gas (LPG) is given as follows (Okori, 1986).

| Component | Ethane | Propane | Isobutane | n-butane | Isopentane |
|-----------|--------|---------|-----------|----------|------------|
| Formula | $C_2H_6$ | $C_3H_8$ | $C_4H_{10}$ | $C_4H_{10}$ | $C_5H_{12}$ |
| Mole % | 0.081 | 32.825 | 24.204 | 42.291 | 0.599 |

Determine the flue gas composition, assuming complete combustion with an air/fuel ratio of 30 to 1.

## Answer

Before equations (24) to (28) above can be used, the formula for LPG has to be expressed in the form $C_kH_r$. This is done remembering that the number of atoms, or moles, of any element in the LPG is equal to $\Sigma x_i C_{jj}$ where $x_i$ is the mole fraction of component, i, and $C_{jj}$ are the atoms, or moles, of an element j in that component. Thus

$$atoms\ of\ C\ in\ LPG = \sum_j x_i C_{ij}$$

$$= 0.00081 \times 2 + 0.32825 \times 3 + 0.24204 \times 4 + 0.42291 \times 4 + 0.00599 \times 5 = 3.676 \tag{1}$$

$$atoms\ of\ H\ in\ LPG = \sum_j x_i C_{ij}$$

$$= 0.00081 \times 6 + 0.32825 \times 8 + 0.24204 \times 10 + 0.42291 \times 10 + 0.00599 \times 12 = 9.35 \tag{2}$$

Hence the molecular formula of this LPG is $C_{3.676}H_{9.35}$. Note also that the fraction of oxygen in air is 0.21 and that, $p = q = m = 0$. We can code the combustion reactants and products as

236

| Component | LPG | $O_2$ | $CO_2$ | $H_2O$ | $N_2$ |
|-----------|-----|-------|--------|--------|-------|
| Code | 1 | 2 | 3 | 4 | 5 |

Using equations (24) to (28) we get the flue gas composition as:

$$x_{34} = x_{H_2O} = \frac{2r}{(r + 4\alpha)} = \frac{2 \times 9.35}{(9.35 + 4 \times 30)} = 0.14457 \quad Ans$$

From (25)

$$x_{33} = x_{CO_2} = \frac{4k}{(r + 4\alpha)} = \frac{4 \times 3.676}{(9.35 + 4 \times 30)} = 0.11368 \quad Ans$$

From (27)

$$x_{35} = x_{N_2} = \frac{2(2\alpha \, x_{25})}{(r + 4\alpha)} = \frac{2(2 \times 30 \times 0.79)}{(9.35 + 4 \times 30)} = 0.73290 \quad Ans$$

From (28)

$$x_{32} = x_{O_2} = \frac{-4k - r + 4\alpha \, x_{22}}{(r + 4\alpha)} = \frac{-4 \times 3.676 - 9.35 + 4 \times 30 \times 0.21}{(9.35 + 4 \times 30)} = 0.00886 \quad Ans$$

As a check,

$$x_{32} + x_{33} + x_{34} + x_{35} = 0.00886 + 0.11368 + 0.14457 + 0.73290 = 1.00001 \; checks$$

If $F_1 = 1$ kmol/h, then $F_2 = 30$ kmol/h and, from equation (23)

$$F_3 = \left(\frac{2p + r + 2q + 4\alpha}{4}\right) F_1 = \left(\frac{9.35 + 4 \times 30}{4}\right) x1 = 32.338 \; kmol \, / \, h$$

These results may be summarized in the Table below as

| Component | $O_2$ | $CO_2$ | $H_2O$ | $N_2$ |
|-----------|-------|--------|--------|-------|
| Mole fraction | 0.00886 | 0.11368 | 0.14457 | 0.73290 |

### 5.2.2.2: Material Balances across a System of Multiple Process Units

The material balances treated so far, have been for single process units. It is the combination of these single process units that make it possible to manufacture or process a product or a range of products. It is, also, this combination that constitutes what are generally referred to as chemical process plants.

It is not difficult to imagine that if material and energy balances can be done for single process units, consideration of the system as a whole may be more efficient and more considerate of the effect and contribution of each process unit to thye effecticeness of the manufacturing process.

To develop the material balances for this system of multiple single
process units, it is usual to group similar process units as modules.
Since most manufacturing processes involve storage, reaction or
synthesis, purifications by separation and product storage, it is usual to
define standard modules in terms of these. The material balance
expressions, already derived for these as single process units are then
combined as they occur in the process and as many times as they .occur,
taking care to have consistent definition .of units. A simple example is
illustrated in Example 5.13 below.

### Illustrative Example 5.13

A waste lime product is to be neutralised, in a process, with 1M $H_2SO_4$
to produce $CaSO_4$ which will be roasted to produce $CaSO_4.\frac{1}{2}H_2O$
(plaster of Paris (POP). The process sequence is shown below. An
analysis of the waste lime shows that it consists of : $Ca(OH)_2$, 61.6 %;
$CaCO_3$, 35.74 %; $H_2O$ 2.39 % and carbon. 0.27 %. Obtain the material
balance for this process.

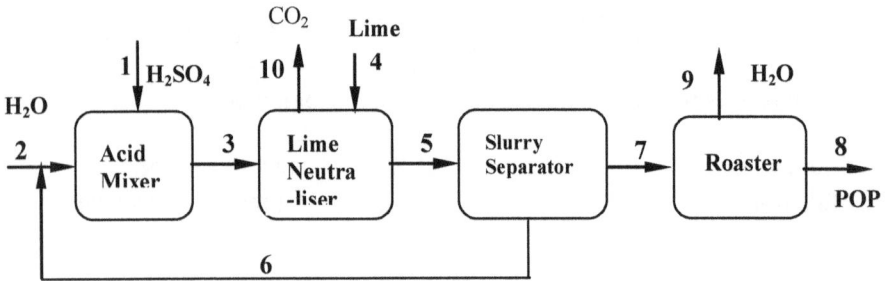

### Answer

The first step is to write out the equations for the reactions taking place,
making sure the equations are balanced. The reactions taking place and
their corresponding equations are shown below.

| | Reaction | Reaction Equations |
|---|---|---|
| i | Neutralisation | $Ca(OH)_2 + H_2SO_4 = CaSO_4 + 2H_2O$ |
| | | $CaCO_3 + H_2SO_4 = CaSO_4 + CO_2 + H_2O$ |
| ii | Roasting | $CaSO_4 + 2H_2O = CaSO_4.2H_2O$ |

238

$$CaSO_4 . 2H_2O = CaSO_4 . \frac{1}{2}H_2O + 1\frac{1}{2}H_2O$$

The next step is to code the streams, process units and components.

| Component | Code | Stream | Code | Process Unit | Code |
|---|---|---|---|---|---|
| $Ca(OH)_2$ | 1 | Inlet $H_2SO_4$ | $F_1$ | Acid mixer | 1 |
| $CaCO_3$ | 2 | Inlet water | $F_2$ | Lime neutralizer | 2 |
| $H_2SO_4$ | 3 | Mixed acid | $F_3$ | Slurry separator | 3 |
| $H_2O$ | 4 | Lime | $F_4$ | Roaster | 4 |
| $CaSO_4$ | 5 | Neutralised lime | $F_5$ | | |
| C | 6 | Supernatant water | $F_6$ | | |
| $CaSO_4.\frac{1}{2}H_2O$ | 7 | Conc solids | $F_7$ | | |
| $CO_2$ | 8 | POP | $F_8$ | | |
| | | Evaporated water | $F_9$ | | |
| | | $CO_2$ | $F_{10}$ | | |

The third step is to carry out material balances around each process unit. Thus,

**For the Acid Mixer**

<u>Material balance equations</u>

| Compound | Code | Material balance equation | |
|---|---|---|---|
| $H_2SO_4$ | 1 | $F_1 x_{13} = F_3 x_{33}$ | (1) |
| water | 2 | $F_1 x_{14} + F_2 x_{24} = F_3 x_{34}$ | (2) |
| Overall | | $F_1 + F_2 = F_3$ | (3) |

<u>Mole fraction constraints</u>

| | | |
|---|---|---|
| Acid Feed | $x_{13} + x_{14} = 1$ | (4) |
| Acid Solution | $x_{33} + x_{34} = 1$ | (5) |
| Water Feed | $x_{24} = 1$ | (6) |

<u>Equipment constraints</u>

$$\alpha = \frac{F_2}{F_1} \ (water \ to \ concentrated \ acid \ ratio \tag{7}$$

## Number of variables

$$N_V = 9 \ (3 \ streams + 5 \ mole \ fractions + 1 \ equipment \ constraint) \tag{8}$$

## Design/decision variables

$$N_D = N_V - N_E = 9 - 6 = 3 \tag{9}$$

Let these be $\alpha$, $F_3$, and $x_{33}$.

## Solution

From equation (7)

$$F_2 = \alpha F_1 \tag{10}$$

From equations (3) and (10)

$$F_3 = F_1 + \alpha F_1 = (1 + \alpha) F_1 \tag{11}$$

From equation (1)

$$x_{33} = \frac{x_{13} F_1}{F_3} \tag{12}$$

From equation (5)

$$x_{34} = 1 - x_{33} \tag{13}$$

## For the Lime Neutraliser

The reactions are

$$Ca(OH)_2 + H_2SO_4 = CaSO_4 + 2 H_2O$$

$$CaCO_3 + H_2SO_4 = CaSO_4 + CO_2 + H_2O$$

Complete reaction is assumed.

## Material balance equations

| Compound | Code | Element | Element balance equation | |
|----------|------|---------|--------------------------|---|
| Ca(OH)$_2$ | 1 | H | $2 F_4 x_{41} + 2 F_3 x_{33} + 2 F_4 x_{44}$ $+ 2 F_3 x_{34} = 2 F_5 x_{54}$ | (14) |
| CaCO$_3$ | 2 | S | $F_3 x_{33} = F_5 x_{55}$ | (15) |
| H$_2$SO$_4$ | 3 | Ca | $F_4 x_{41} + F_4 x_{42} = F_5 x_{55}$ | (16) |

| | | | | |
|---|---|---|---|---|
| $H_2O$ | 4 | O | $2F_4 x_{41} + 3F_4 x_{42} + F_4 x_{44} + 4F_3 x_{33}$ $+ F_3 x_{34} = 4F_5 x_{55} + F_5 x_{54} + 2F_{10} x_{108}$ | (17) |
| $CaSO_4$ | 5 | Bound carbon | $F_4 x_{42} = F_{10} x_{108}$ | (18a) |
| $C$ | 6 | Free carbon | $F_4 x_{46} = F_5 x_{56}$ | (18b) |
| $CO_2$ | 8 | | | |
| Overall | | | $F_3 + F_4 = F_5 + F_{10}$ | (19) |

## Mole fraction constraints

Acid Solution $\quad x_{33} + x_{34} = 1 \quad$ (20)

Lime Feed $\quad x_{41} + x_{42} + x_{44} + x_{46} = 1 \quad$ (21)

Neutralised Lime Stream $\quad x_{54} + x_{55} + x_{56} = 1 \quad$ (22)

$CO_2$ Stream $\quad x_{108} = 1 \quad$ (23)

## Equipment constraints

$$K_1 = C_1 \quad for \quad neutralisation\ of\ strong\ acid\ \&\ base \quad (24)$$
$$K_2 = C_2 \quad for \quad neutralisation\ of\ weak\ acid\ \&\ base \quad (25)$$
$$\beta = \frac{F_3}{F_4} \ (mixed\ acid\ to\ lime\ ratio) \quad (26)$$

## Number of variables

$$N_V = 17 \ (4\ streams + 10\ mole\ fractions + 3\ equipment\ constraints) \quad (27)$$

## Design/decision variables

$$N_D = N_V - N_E = 17 - 13 = 4 \quad (28)$$

Let these be $F_5$, $x_{54}$, $x_{55}$, $x_{56}$.

## Solution

From equation (19)
$$F_5 = F_3 + F_4 - F_{10} \quad (29)$$

From equation (15)
$$x_{55} = \frac{F_3 x_{33}}{F_5} \quad (30)$$

From equation (16)

$$x_{55} = \frac{F_4 (x_{41} + x_{42})}{F_5} \tag{31}$$

From equation (18a)

$$F_{10} = \frac{F_4 x_{42}}{x_{108}} \tag{32}$$

From equation (18b)

$$x_{56} = \frac{F_4 x_{46}}{F_5} \tag{33}$$

From equation (14) and (20)

$$x_{54} = \frac{2 F_4 x_{41} + 2 F_3 x_{33} + 2 F_4 x_4 + 2 F_3 x_{34}}{2 F_5} = \frac{F_4 (x_{41} + x_{44}) + F_3}{F_5} \tag{34}$$

From equation (17)

$$x_{54} = \frac{F_4 \left(2 x_{41} + 3 x_{42} + x_{44}\right) + F_3 \left(4 x_{33} + x_{34}\right) - 4 F_5 x_{55} - 2 F_{10} x_{108}}{F_5} \tag{34}$$

From equations (26)

$$F_3 = \beta F_4 \tag{35}$$

**For the Slurry Concentrator**

Assume complete reaction. Let $\alpha$ be the ratio of moles of water to those of the $CaSO_4$ stream

| Compound | Code | Material balance equation | |
|----------|------|---------------------------|------|
| Ca(OH)$_2$ | 1 | | |
| CaCO$_3$ | 2 | | |
| H$_2$SO$_4$ | 3 | | |
| H$_2$O | 4 | $F_5 x_{54} = F_7 x_{74} + F_6 x_{64}$ | (36) |
| CaSO$_4$ | 5 | $F_5 x_{55} = F_7 x_{75}$ | (37) |
| C | 6 | $F_5 x_{56} = F_7 x_{76} + F_6 x_{66}$ | (38) |
| Overall | | $F_5 = F_7 + F_6$ | (39) |

Mole fraction constraints

Neutralised Lime Feed $\quad x_{54} + x_{55} + x_{56} = 1 \tag{40}$

Overflow stream $\quad x_{64} + x_{66} = 1 \tag{41}$

242

| Underflow stream | $x_{74} + x_{75} + x_{76} = 1$ | (42) |
|---|---|---|

## Equipment constraints

$$x_{76} = 0.001 \tag{43}$$

## Number of variables

$$N_V = 12\,(3 \; streams + 8 \; mole \; fractions + 1 \; equipment \; constraint) \tag{44}$$

## Design/decision variables

$$N_D = N_V - N_E = 12 - 8 = 4 \tag{45}$$

Let these be $F_7$, $x_{66}$, $x_{75}$, and $x_{76}$.

## Solution

From equation (39)

$$F_7 = F_5 - F_6 \tag{46}$$

From equation (37)

$$x_{75} = \frac{F_5 \, x_{55}}{F_7} \tag{47}$$

From equation (38)

$$x_{66} = \frac{F_5 \, x_{56} - F_7 \, x_{76}}{F_6} \tag{48}$$

From equation (41)

$$x_{64} = 1 - x_{66} \tag{49}$$

From equation (36)

$$x_{74} = \frac{F_5 \, x_{54} - F_6 \, x_{64}}{F_7} \tag{50}$$

## For the Roaster

There is some rearrangement of molecules, hence the need for an element balance

## Material balance equations

| Component | Code | Element | Element balance equation |
|---|---|---|---|
| $Ca(OH)_2$ | 1 | | |
| $CaCO_3$ | 2 | | |

243

| $H_2SO_4$ | 3 | H | $2 F_7 x_{74} = 2 F_9 x_{94} + F_8 x_{87}$ | (51) |
|---|---|---|---|---|
| $H_2O$ | 4 | S | $F_7 x_{75} = F_8 x_{87}$ | (52) |
| $CaSO_4$ | 5 | Ca | $F_7 x_{75} = F_8 x_{87}$ | (53) |
| C | 6 | O | $F_7 x_{74} + 4 F_7 x_{75} = 5 F_8 x_{87} + F_9 x_{94}$ | (54) |
| $CaSO_4.\frac{1}{2}H_2O$ | 7 | C | $F_7 x_{76} = F_8 x_{86}$ | (55) |

## Mole fraction constraints

| Feed Stream | $x_{74} + x_{75} + x_{76} = 1$ | (56) |
|---|---|---|
| Product Stream | $x_{87} + x_{86} = 1$ | (57) |
| Water vapour | $x_{94} = 1$ | (58) |

## Equipment constraints

$$x_{86} = 0.001 \tag{59}$$

## Number of variables

$$N_V = 10 \, (3 \; streams + 6 \; mole \; fractions + 1 \; equipment \; constraint s) \tag{60}$$

## Design/decision variables

$$N_D = N_V - N_E = 10 - 8 = 2 \tag{61}$$

Let these be $F_8$, $x_{87}$.

## Solution

From equations (55)

$$F_8 = \frac{F_7 x_{76}}{x_{86}} \tag{62}$$

From equation (52)

$$x_{87} = \frac{F_7 x_{75}}{F_8} \tag{63}$$

From equation (54)

$$F_8 = \frac{F_7 x_{74} + 4 F_7 x_{75} - F_9}{5 x_{87}} \tag{64}$$

As a check

$$F_8 = F_7 - F_9 \qquad (65)$$

Equations (10) to (12), (29) to (35), (46) to (50), (62) to (65) constitute the solution algorithm for the problem.

Suppose an analysis of the waste lime feed is $Ca(OH)_2$, 61.6 %; $CaCO_3$, 35.74 %; $H_2O$, 2.39 %; Carbon, 0.27 % and 98% , 36.7N $H_2SO_4$ (S.G = 1.8361) is to be mixed to 2.17N.(S.G = 1.0661), the starting compositions for the algorithm are evaluated as shown below.

**Evaluation of Known Feed and Stream Conditions**

**To determine $F_1$.**

The reactions are

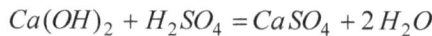

$$CaCO_3 + H_2SO_4 = CaSO_4 + CO_2 + H_2O$$
$$Ca(OH)_2 + H_2SO_4 = CaSO_4 + 2H_2O$$

Thus

$100 \ kg \ CaCO_3$ reacts with $98 \ kg \ H_2SO_4$ to form $136 \ kg \ CaSO_4$
$$+ 44 \ kg \ CO_2 + 18 \ kg \ H_2O$$

$357.4 \ kg \ CaCO_3$ will react with $\dfrac{98}{100} x \ 357.4 = 350.252 \ kg \ H_2SO_4$

to form $\dfrac{136}{100} x \ 357.4 = 486.064 \ kg \ CaSO_4 + \dfrac{44}{100} x \ 357.4 = 157.256 \ kg \ CO_2$

$$+ \dfrac{18}{100} x \ 357.4 = 64.332 \ kg \ H_2O$$

Similarly

$74 \ kg \ Ca(OH)_2$ reacts with $98 \ kg \ H_2SO_4$ to form $136 \ kg \ CaSO_4 + 36 \ kg \ H_2O$

$616 \ kg \ Ca(OH)_2$ will react with $\dfrac{98}{74} x \ 616 = 815.784 \ kg \ H_2SO_4$

to form $\dfrac{136}{74} x \ 616 = 1132.108 \ kg \ CaSO_4 + \dfrac{36}{74} x \ 616 = 299.676 \ kg \ H_2O$

Thus, 350.252 + 815.784 = 1166.036 kg/h or 11.898 kmol/h or 23.796 kg equivalents of $H_2SO_4$ would be required.

Since 1 litre of 36.7 N $H_2SO_4$ contains 36.7 x 49 = 1798.3 gram of

$H_2SO_4$, 1,166,036 grams would be provided by 1,166,036/1798.3 = 648.410 litres. This is $F_1$ in litres. The weight of the 36.7 N $H_2SO_4$, using the given specific gravity, is 648,410 x 1.8361/1000 = 1190.546 kg/h. This is $F_1$ in kg/h.

Similarly, since 1 litre of 2.17 N $H_2SO_4$ contains 2.17 x 49 = 106.33 gram of $H_2SO_4$, then 1,166,036 grams/h of $H_2SO_4$ would require 1,166,036 /106.33 = 10966.200 litres of 2.17 N $H_2SO_4$. This is $F_3$ in litres. The weight of this 2.17 N $H_2SO_4$, using the given specific gravity of 1.0661, is 10,966,200 x 1.0661/1000 = 11,697.646 kg/h. This is $F_3$ in kg/h.

## To determine $\alpha = F_2/F_1$

Since the mixed acid is 11,697.646 kg/h from 1190.546 kg/h of concentrated acid, amount of water required is 11,697.646 − 1190.546 = 10,507.1 kg/h. This is $F_2$ in kg/h.

The compositions of streams $F_1$, $F_2$ and $F_3$ are then calculated as shown below.

**Table 1:  Molar Composition of 36.7 N $H_2SO_4$ Feed**

| Component | Actual Weight, kg/h | kmoles/h | Mole Fraction |
|---|---|---|---|
| $H_2O$ | 0.02 x 1190.546 = 23.811 | $\dfrac{23.811}{18} = 1.323$ | $x_{14} = 0.1000$ |
| $H_2SO_4$ | 0.98 x 1190.546 = 1166.735 | $\dfrac{1166.735}{98} = 11.906$ | $x_{13} = 0.9000$ |
| Totals | 1,190.546 | $F_1 = 13.229$ | 1.0000 |

**Table 2: Molar Composition of $H_2O$ Feed ($F_2$)**
**Basis: 1000 kg/h of Lime Feed**

| Component | Actual Weight, kg/h | kmoles/h | Mole Fraction |
|---|---|---|---|
| $H_2O$ | 10,507.1 | $\dfrac{10,507.1}{18} = 583.728 = F_2$ | $x_{24} = 1.0000$ |

Hence
$$\alpha = \frac{F_2}{F_1} = \frac{583.728}{13.229} = 44.125 \qquad (66)$$

### Table 3: Molar Composition of Mixed Acid ($F_3$)
### Basis: 1000 kg/h of Lime Feed

| Component | Weight, kg/h | kmol/h | Mole fraction |
|---|---|---|---|
| $H_2O$ | $10,507.1 + 23.811 = 10,530.911$ | $\frac{10530.911}{18} = 585.051$ | $x_{34} = 0.9800$ |
| $H_2SO_4$ | 1166.735 | $\frac{1166.735}{98} = 11.905$ | $x_{33} = 0.0200$ |
| Totals | 11,697.646 | $F_3 = 596.956$ | 1.0000 |

From equation (11)
$$F_3 = (1+\alpha)F_1 = (1+44.125) \times 13.229 = 596.959 \; kmol \, / \, h \qquad (67)$$
Checks.

### Table 4:  Molar Composition of Waste Lime Feed ($F_4$)
### Basis 1000 kg/h

| Component | Weight Fraction | Actual weight, kg/h | Molecular weight | Moles/h |
|---|---|---|---|---|
| $Ca(OH)_2$ | 0.6160 | 616.0 | 74 | 8.324 |
| $CaCO_3$ | 0.3574 | 357.4 | 100 | 3.574 |
| $H_2O$ | 0.0239 | 23.9 | 18 | 1.328 |
| Carbon | 0.0027 | 2.7 | 12 | 0.225 |
| Total | 1.0000 | 1000.0 | | $F_4 = 13.451$ |

### Table 5:  Mole Fraction Composition of Waste Lime Feed ($F_4$)
### Basis 1000 kg/h

| Component | Moles/h | Mole Fraction | Symbol |
|---|---|---|---|
| $Ca(OH)_2$ | 8.324 | 0.6188 | $x_{41}$ |
| $CaCO_3$ | 3.574 | 0.2657 | $x_{42}$ |
| $H_2O$ | 1.328 | 0.0987 | $x_{44}$ |
| Carbon | 0.225 | 0.0167 | $x_{46}$ |
| Total | $F_4 = 13.451$ | 0.9999 | |

247

## To determine the Composition of the $CO_2$ Stream arising from Lime Neutralisation

The reaction between $CaCO_3$ and $H_2SO_4$ is the only one in which $CO_2$ is produced. Thus:

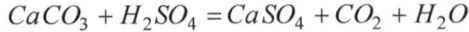

$$CaCO_3 + H_2SO_4 = CaSO_4 + CO_2 + H_2O$$

1000 kg/h of lime feed contains 35.74 %, that is, 357.4 kg/h, of $CaCO_3$. This will react with $H_2SO_4$, according to the above equation, to produce 157.256 kg/h, or 3.574 kmols, of $CO_2$.

**Table 6: Composition of the $CO_2$ Stream ($F_{10}$)**
**Basis: 1000 kg/h Feed Lime**

| Component | kmoles/h | Mole Fraction | Symbol |
|-----------|----------|---------------|--------|
| $CO_2$ | 3.574 | 1.0000 | $X_{108}$ |
| Totals | 3.574= $F_{10}$ | 1.0000 | |

**Table 7: Mole Fraction Composition of Neutralised Lime Product ($F_5$)**
**Basis 1000 kg/h of Lime Feed**

| Component | Moles/h | Mole Fraction | Symbol |
|-----------|---------|---------------|--------|
| $CaSO_4$ | 11.898 | 0.0192 | $X_{55}$ |
| $H_2O$ from reaction | 20.223 | 0.9804 | $X_{54}$ |
| $H_2O$ from mixed acid | 585.051 | | |
| $H_2O$ from lime feed | 1.328 | | |
| Carbon | 0.225 | 0.0004 | $X_{56}$ |
| **Total** | $F_5 = 618.725$ | 1.0000 | |

**Table 8: Mole Fraction Composition of Concentrated $CaSO_4$ Product ($F_7$)**
**Basis 1000 kg/h of Lime Feed**

| Component | Moles/h | Mole Fraction |
|-----------|---------|---------------|
| $CaSO_4$ | 11.898 | $x_{75} = \dfrac{11.898}{11.898(1+\alpha)+0.225\theta}$ |

| $H_2O$ | $11.898\alpha$ | $x_{74} = \dfrac{11.898\,\alpha}{11.898\,(1+\alpha)+0.225\theta}$ |
|--------|----------------|------------------------------------------------------------------|
| Carbon | $0.225\theta$ | $x_{76} = \dfrac{0.225\theta}{11.898\,(1+\alpha)+0.225\theta}$ |
| **Total** | $F_7 = 11.898(1+\alpha)+0.225\theta$ | |

$\alpha$ is the ratio of water to $CaSO_4$ in the concentrated $CaSO_4$ stream.
$\theta$ is the fraction of the original carbon feed left in the $CaSO_4$ stream.

Having, thus, estimated known compositions from given data, we can proceed to evaluate the rest of the algorithm.

**For the Acid Mixer**

From equation (12)

$$x_{33} = \frac{x_{13}\,F_1}{F_3} = \frac{0.90 \times 13.229}{596.956} = 0.0200 \qquad (68)$$

From equation (13)

$$x_{34} = 1 - x_{33} = 1 - 0.0200 = 0.9800 \qquad (69)$$

**For the Lime Neutraliser**

From equation (30)

$$x_{55} = \frac{F_3\,x_{33}}{F_5} = \frac{596.956 \times 0.02}{618.725} = 0.0193 \qquad (70)$$

From equation (32)

$$F_{10} = \frac{F_4\,x_{42}}{x_{108}} = \frac{13.451 \times 0.2657}{1} = 3.574 \qquad (71)$$

From equation (33)

$$x_{56} = \frac{F_4\,x_{46}}{F_5} = \frac{13.451 \times 0.0167}{618.725} = 0.0004 \qquad (72)$$

From equation (34)

$$x_{54} = \frac{F_4\,(x_{41}+x_{44})+F_3}{F_5} = \frac{13.451(0.6188+0.0987)+596.956}{618.725} = 0.9804 \qquad (73)$$

From equation (26)

$$\beta = \frac{F_3}{F_4} = \frac{596.956}{13.451} = 44.380 \tag{74}$$

## For the Slurry Concentrator

Let $\alpha = 1.5$. Also let 90 % of carbon be removed from the $CaSO_4$ stream. That is, $\theta = 0.1$

From Table 8
$$F_7 = 11.898 \, x \, 2.5 + 0.225 \, x \, 0.1 = 29.767 \tag{75}$$

From equation (46) and Table 7
$$F_6 = F_5 - F_7 = 618.725 - 35.717 = 588.958 \tag{76}$$

From equation (47)
$$x_{75} = \frac{F_5 \, x_{55}}{F_7} = \frac{618.725 \, x \, 0.0192}{29.767} = 0.3991 \tag{77}$$

From Table 8
$$x_{75} = \frac{11.898}{11.898 \, (1+\alpha) + 0.0225} = \frac{11.898}{11.898 \, x \, 2.5 + 0.0225} = 0.3997 \tag{77a}$$

From Table 8
$$x_{76} = \frac{0.0225}{11.898 \, (1+\alpha) + 0.0225} = \frac{0.0225}{11.898 \, x \, 2.5 + 0.0225} = 0.0008 \tag{78}$$

From equation (48)
$$x_{66} = \frac{F_5 \, x_{56} - F_7 \, x_{76}}{F_6} = \frac{618.725 \, x \, 0.0004 - 29.767 \, x \, 0.0008}{588.958} = 0.0004 \tag{79}$$

From equation (49)
$$x_{64} = 1 - x_{66} = 1 - 0.0004 \approx 0.9996 \tag{80}$$

From equation (50)
$$x_{74} = \frac{F_5 \, x_{54} - F_6 \, x_{64}}{F_7} = \frac{618.725 \, x \, 0.9804 - 588.958 \, x \, 0.9996}{29.767} = 0.6005 \tag{81}$$

From Table 8
$$x_{74} = \frac{11.898 \, \alpha}{11.898 \, (1+\alpha) + 0.0225} = \frac{11.898 \, x \, 1.5}{11.898 \, x \, 2.5 + 0.0225} = 0.5995 \tag{81a}$$

## For the Roaster

The reaction is

$$CaSO_4 + 2H_2O = CaSO_4 \cdot \tfrac{1}{2}H_2O + \tfrac{3}{2}H_2O$$

Assuming 100 % conversion, 11.898 kmol/h of $CaSO_4$ will form 11.898 kmol/h of $CaSO_4 \cdot \tfrac{1}{2}H_2O$ and 1.5 x 11.898 = 17.847 kmol/h of $H_2O$

Thus, $F_8 = 11.898$ kmol/h. $F_9 = 17.847$ kmol/h

From equations (62)

$$x_{86} = \frac{F_7\, x_{76}}{F_8} = \frac{29.767 \ x \ 0.0008}{11.898} = 0.0020 \qquad (82)$$

From equation (62)

$$x_{87} = \frac{F_7\, x_{75}}{F_8} = \frac{29.767 \ x \ 0.3991}{11.898} = 0.9985 \qquad (83)$$

As a check, from (65)

$$F_8 = F_7 - F_9 = 29.767 - 17.847 = 11.92 \qquad (84)$$

This is only 0.18 % in error; checks.

## Table 9:  Summary of Stream Mole Fraction Compositions

### Stream 1: Concentrated Acid (36.7N $H_2SO_4$) Feed, 13.229 kmol/h

| Component | $H_2SO_4$ | $H_2O$ | Total |
|---|---|---|---|
| Mole Fraction | 0.9000 | 0.1000 | 1.0000 |

### Stream 2: Fresh Water ($H_2O$) Feed, Initial Feed 583.278 kmol/h Removal of 5.68 kmol/h from Recycle Stream

| Component | $H_2O$ | Total |
|---|---|---|
| Mole Fraction | 1.0000 | 1.0000 |

### Stream 3: Mixed Acid (2.17N $H_2SO_4$) Feed, 596.956 kmol/h

| Component | $H_2SO_4$ | $H_2O$ | Total |
|---|---|---|---|
| Mole Fraction | 0.0200 | 0.9800 | 1.0000 |

### Stream 4: Fresh Lime Feed, 13.451 kmol/h

| Component | $Ca(OH)_2$ | $CaCO_3$ | $H_2O$ | C | Total |
|---|---|---|---|---|---|

| Mole Fraction | 0.6188 | 0.2657 | 0.0987 | 0.0167 | 0.9999 |

### Stream 5: Neutralised Lime Feed, 618.725 kmol/h

| Component | $CaSO_4$ | $H_2O$ | C | Total |
|---|---|---|---|---|
| Mole Fraction | 0.0192 | 0.9804 | 0.0004 | 1.0000 |

### Stream 6: Overflow from Slurry Concentrator, 588.958 kmol/h

| Component | $H_2O$ | C | Total |
|---|---|---|---|
| Mole Fraction | 0.9996 | 0.0004 | 1.0000 |

### Stream 7: Concentrated Underflow Slurry Feed, 29.767 kmol/h

| Component | $CaSO_4$ | $H_2O$ | C | Total |
|---|---|---|---|---|
| Mole Fraction | 0.3997 | 0.5995 | 0.0008 | 1.0000 |

### Stream 8: Plaster of Paris (POP) Product, 11.898 kmol/h

| Component | $CaSO_4.\frac{1}{2} H_2O$ | C | Total |
|---|---|---|---|
| Mole Fraction | 0.9985 | 0.0020 | 1.0005 |

### Stream 9: Evaporated Water from Roaster, 17.847 kmol/h

| Component | $H_2O$ | Total |
|---|---|---|
| Mole Fraction | 1.0000 | 1.0000 |

### Stream 10: $CO_2$ from $CaCO_3$ Neutralisation Reaction, 3.574 kmol/h

| Component | $CO_2$ | Total |
|---|---|---|
| Mole Fraction | 1.0000 | 1.0000 |

## 5.2.3: The Split Fraction Method

This method has been described in detail, together with a computer program (MASSBAL) for its implementation, by Coulson, Richardson & Sinnott (1983). Only the basic procedure, sufficient for general purpose use, is described here. The method is applied in four stages and

252

these may be summarised as follows:

1. Develop the information flow diagram
2. Develop the mass balance equations in terms of split fraction coefficients and fresh feds
3. Estimate the split fraction coefficients and fresh feeds
4. Solve the resulting simulataneous equations

## Stage 1: Developing the Information Flow Diagram

The information flow diagram is a diagram in which blocks, connected by arrows, are used to represent caalculationh modules in the process (diagraph). Such calculation modules relate outlet to inlet flows.

Only units, in which there is a change of composition, temperature or pressure, are shown. This ensures that the flow of information, from one set of calculations to the next, is, clearly, illustrated.

## Stage 2: Developing the Mass Balance Equations

If we define
$i$ = unit number
$k$ = component number
$\lambda_{ik}$ = flow of component $k$ into unit $i$
$\alpha_{jik}$ = split fraction coefficient = fraction of total flow of component $k$ from unit $i$ which leaves unit $i$ to enter unit $j$
$g_{iok}$ = any fresh feed of component $k$ entering unit $i$ or which flows in from outside the system (unit 0)
we can specify a calculation module in an information flow diagram as shown in the diagram below.

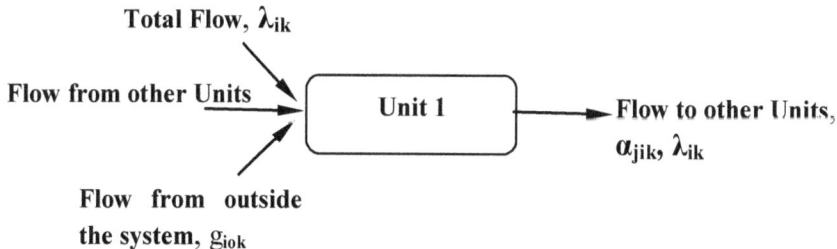

**Total Flow, $\lambda_{ik}$**

**Flow from other Units**  →  | **Unit 1** |  → **Flow to other Units,**
$\alpha_{jik}, \lambda_{ik}$

**Flow from outside the system, $g_{iok}$**

The flow of any component, $k$, from unit $i$ to unit $j$ is given by $\alpha_{jik}\,\lambda_{ik}$. Consider, now, the mass balance on a system made up of two calculation modules as shown below.

253

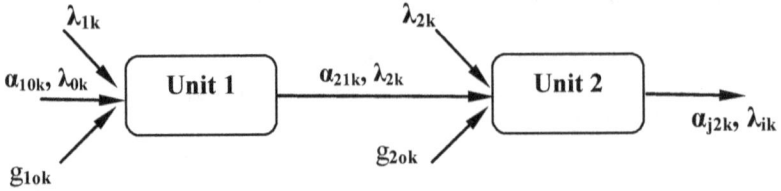

Mass balance over unit 1

$$\lambda_{1k} = g_{10k} + \alpha_{10k} \cdot \lambda_{0k} \quad for\ all\ k \qquad (5.67)$$

Or

$$\lambda_{1k} - \alpha_{10k} \cdot \lambda_{0k} = g_{10k} \qquad (5.67a)$$

Mass balance over unit 2

$$\lambda_{2k} = g_{20k} + \alpha_{21k} \cdot \lambda_{1k} \quad for\ all\ k \qquad (5.68)$$

Or

$$\lambda_{2k} - \alpha_{21k} \cdot \lambda_{1k} = g_{20k} \qquad (5.68a)$$

The equations to be solved are equations (5.67a) and (5.68a). In matrix notation, using the $\lambda_{ik}$ as the independent variable, we get

$$\begin{pmatrix} -\alpha_{10k} & 1 & 0 \\ 0 & -\alpha_{21k} & 1 \end{pmatrix} \cdot \begin{pmatrix} \lambda_{0k} \\ \lambda_{1k} \\ \lambda_{2k} \end{pmatrix} = \begin{pmatrix} g_{10k} \\ g_{20k} \end{pmatrix} \qquad (5.69)$$

For a system of two components. $k = 2$ and the matrices to be solved are, for k= 1:

$$\begin{pmatrix} -\alpha_{101} & 1 & 0 \\ 0 & -\alpha_{211} & 1 \end{pmatrix} \cdot \begin{pmatrix} \lambda_{01} \\ \lambda_{11} \\ \lambda_{21} \end{pmatrix} = \begin{pmatrix} g_{101} \\ g_{201} \end{pmatrix} \qquad (5.70)$$

and, for k=2:

$$\begin{pmatrix} -\alpha_{102} & 1 & 0 \\ 0 & -\alpha_{212} & 1 \end{pmatrix} \cdot \begin{pmatrix} \lambda_{02} \\ \lambda_{12} \\ \lambda_{22} \end{pmatrix} = \begin{pmatrix} g_{102} \\ g_{202} \end{pmatrix} \qquad (5.71)$$

### Stage 3: Estimating Split Fraction Coefficients and Fresh Feeds

Coulson, Richardson & Sinnott (1983) give six general guidelines for estimating split fraction coefficients which are summarised below.

1.  Reactors: For chemical reactions, the fractional conversion is the split fraction coefficient.

2. <u>Mixers</u>: The split fraction coefficient is equal to 1 for each component.

3. <u>Stream dividers</u>: Provided there is no change in composition, the split fraction ooefficient is equal to the fractional division of the stream.

4. <u>Absorption, Stripping, Solvent Extraction, Leaching, etc</u>: Since the fraction of component absorbed, stripped, extracted or leached, etc, depends on column design, such as the number of stages, the component solubility, as well as the flowrates of the light and heavy streams, the use of simplified formulae, such as absorption factor method of Kremser, for absorption, is encouraged.

5. <u>Distillation Columns</u>: Here, the split fraction coefficient is given by
$$\frac{x_{sk} . R}{x_{ik}}$$ where $x_{sk}$ is the concentration of component $k$ in stream $s$, $x_{ik}$
is the concentration of component $k$ in the feed stream and R is the fraction of total feed that goes to stream $s$.

6. <u>Equilibrium separators</u>: If the equilibrium constant can be defined as
$$K = \frac{x_{ak}}{x_{bk}}$$ where $x_{ak}$ is concentration of component $k$ in stream $a$ and
$x_{bk}$ is the concentration of component $k$ in stream $b$ then the split
fraction coefficient for stream $a$ is given by $\left(\frac{K}{K-1}\right) . \left(\frac{x_{jk} - x_{bk}}{x_{jk}}\right)$
where $x_{jk}$ is the concentration of component $k$ in the feed stream.

Fresh feeds are estimated directly from the feed specifications or from the product specifications using a mass balance or, in the case of reactions, from stoichiometry.

**Stage 4: Solving tbe Resulting Equations**

The matrices that result are, usually, sparse. Depending on the number of components and calculation modules, these matrices may be solved, for very simple cases, using standard procedures.

For complex.problems, sparse matrix solution routines, such as those of Gunn (Sinnott, 1983) and Thomas (Perry & Green, 1984), which save computational effort and time, are used. To illustrate the method, the problem of Example 5.13 is reworked in Example 5.14 using the Split Fraction Method.

Rework Example 5.13 as **Example 5.14** using the Split Fraction Method.

The process sequence is the same as before and is shown below.

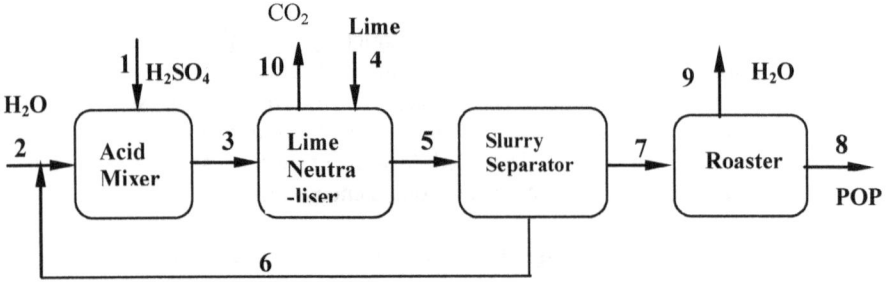

## Stage 1: Developing the Information Flow Diagram

For ease of comparison, let us retain the coding of the streams, process units and components used in the method of constraints.

| Component | Code | Process Unit | Code |
|---|---|---|---|
| $Ca(OH)_2$ | 1 | Acid mixer | 1 |
| $CaCO_3$ | 2 | Lime neutralizer | 2 |
| $H_2SO_4$ | 3 | Slurry separator | 3 |
| $H_2O$ | 4 | Roaster | 4 |
| $CaSO_4$ | 5 | | |
| C | 6 | | |
| $CaSO_4.\frac{1}{2}H_2O$ | 7 | | |
| $CO_2$ | 8 | | |

The information flow diagram is developed as shown below, using the process unit codes above.

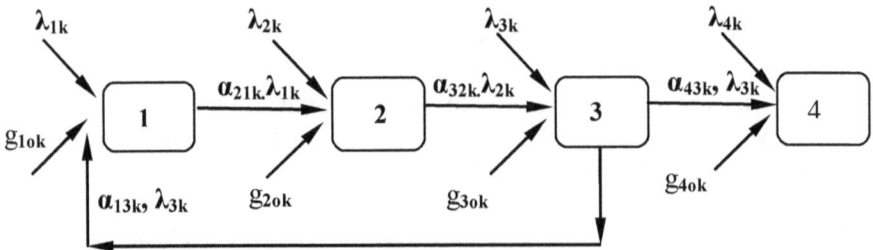

## Stage 2: Developing the Material Balance

The material balance, involving all components, is shown below.

Complete reaction is assumed.

| | | | | |
|---|---|---|---|---|
| Unit 1 | $\lambda_{1k} = g_{10k} + \alpha_{13k}.\lambda_{3k}$ | or | $\lambda_{1k} - \alpha_{13k}.\lambda_{3k} = g_{10k}$ | (1) |
| Unit 2 | $\lambda_{2k} = g_{20k} + \alpha_{21k}.\lambda_{1k}$ | or | $-\alpha_{21k}.\lambda_{1k} + \lambda_{2k} = g_{20k}$ | (2) |
| Unit 3 | $\lambda_{3k} = g_{30k} + \alpha_{32k}.\lambda_{2k}$ | or | $-\alpha_{32k}.\lambda_{2k} + \lambda_{3k} = g_{30k}$ | (3) |
| Unit 4 | $\lambda_{4k} = g_{40k} + \alpha_{43k}.\lambda_{3k}$ | or | $-\alpha_{43k}.\lambda_{3k} + \lambda_{4k} = g_{40k}$ | (4) |

Equations (1) to (4) can be expressed in matrix notation as

$$
\begin{pmatrix}
1 & 0 & -\alpha_{13k} & 0 \\
-\alpha_{21k} & 1 & 0 & 0 \\
0 & -\alpha_{32k} & 1 & 0 \\
0 & 0 & -\alpha_{43k} & 1
\end{pmatrix}
\cdot
\begin{pmatrix}
\lambda_{1k} \\
\lambda_{2k} \\
\lambda_{3k} \\
\lambda_{4k}
\end{pmatrix}
=
\begin{pmatrix}
g_{10k} \\
g_{20k} \\
g_{30k} \\
g_{40k}
\end{pmatrix}
\tag{5}
$$

This matrix can be solved if the values of the split fraction coefficients, $\alpha_{jik}$, and fresh feeds, $g_{i0k}$, for all components, are known.

## Stage 3: Estimating the Split Fraction Coefficients and Fresh Feeds

We can use the calculations of Tables 1 to 8 of the previous example since they do not depend on the process method of carrying out the material balance. Thus,

## Estimation of Fresh Feed of Components, $g_{iok}$

| | Unit 1 | Unit 2 | Unit 3 | Unit 4 |
|---|---|---|---|---|
| Ca(OH)$_2$ (1) | $g_{101} = 0$ | $g_{201} = 8.324$ | $g_{301} = 0$ | $g_{401} = 0$ |
| CaCO$_3$ (2) | $g_{102} = 0$ | $g_{202} = 3.574$ | $g_{302} = 0$ | $g_{402} = 0$ |
| H$_2$SO$_4$ (3) | $g_{103} = 11.906$ | $g_{203} = 0$ | $g_{303} = 0$ | $g_{403} = 0$ |
| H$_2$O (4) | $g_{104} = 1.323$ $+582.405$ $= 583.728$ | $g_{204} = 1.328$ $+20.223$ $= 21.551$ | $g_{304} = 0$ | $g_{404} =$ $17.847$ |
| CaSO$_4$ (5) | $g_{105} = 0$ | $g_{205} = 11.898$ | $g_{305} = 0$ | $g_{405} = 0$ |
| C (6) | $g_{106} = 0$ | $g_{206} = 0.225$ | $g_{306} = 0$ | $g_{406} = 0$ |

257

| $CaSO_4.\frac{1}{2}H_2O$ (7) | $g_{107} = 0$ | $g_{207} = 0$ | $g_{307} = 0$ | $g_{407} = 11.898$ |
|---|---|---|---|---|
| $CO_2$ (8) | $g_{108} = 0$ | $g_{208} = 3.574$ | $g_{308} = 0$ | $g_{408} = 0$ |

## Estimation of Split Fraction Coefficients

Assume, like in Example 5.13, that $\theta = 0.1$ and $\alpha = 1.5$

| | Unit 1 | Unit 2 | Unit 3 | Unit 4 |
|---|---|---|---|---|
| $Ca(OH)_2$ (1) | $\alpha_{131} = 0$ | $\alpha_{211} = 0$ | $\alpha_{321} = 0$ | $\alpha_{431} = 0$ |
| $CaCO_3$ (2) | $\alpha_{132} = 0$ | $\alpha_{212} = 0$ | $\alpha_{322} = 0$ | $\alpha_{432} = 0$ |
| $H_2SO_4$ (3) | $\alpha_{133} = 0$ | $\alpha_{213} = 0.02$ | $\alpha_{323} = 0$ | $\alpha_{433} = 0$ |
| $H_2O$ (4) | $\alpha_{134} = 1$ | $\alpha_{214} = 0.98$ | $\alpha_{324} = 0.9804$ | $\alpha_{434} = 0.5995$ |
| $CaSO_4$ (5) | $\alpha_{135} = 0$ | $\alpha_{215} = 0$ | $\alpha_{325} = 0.0192$ | $\alpha_{435} = 0.3997$ |
| C (6) | $\alpha_{136} = 0$ | $\alpha_{216} = 0$ | $\alpha_{326} = 0.0004$ | $\alpha_{436} = 0.0008$ |
| $CaSO_4.\frac{1}{2}H_2O$ (7) | $\alpha_{137} = 0$ | $\alpha_{217} = 0$ | $\alpha_{327} = 0$ | $\alpha_{437} = 0$ |
| $CO_2$ (8) | $\alpha_{138} = 0$ | $\alpha_{218} = 0$ | $\alpha_{328} = 0$ | $\alpha_{438} = 0$ |

Hence the matrices to be evaluated are

Component 1: $Ca(OH)_2$

$$\begin{pmatrix} 1 & 0 & 0 & 0 \\ 0 & 1 & 0 & 0 \\ 0 & 0 & 1 & 0 \\ 0 & 0 & 0 & 1 \end{pmatrix} \begin{pmatrix} \lambda_{11} \\ \lambda_{21} \\ \lambda_{31} \\ \lambda_{41} \end{pmatrix} = \begin{pmatrix} 0 \\ 8.324 \\ 0 \\ 0 \end{pmatrix} \qquad (6)$$

The solution is $\lambda_{11} = 0$; $\lambda_{21} = 8.324$ kmol/h; $\lambda_{31} = 0$; $\lambda_{41} = 0$    (7)

Component 2: $CaCO_3$

$$\begin{pmatrix} 1 & 0 & 0 & 0 \\ 0 & 1 & 0 & 0 \\ 0 & 0 & 1 & 0 \\ 0 & 0 & 0 & 1 \end{pmatrix} \begin{pmatrix} \lambda_{12} \\ \lambda_{22} \\ \lambda_{32} \\ \lambda_{42} \end{pmatrix} = \begin{pmatrix} 0 \\ 3.574 \\ 0 \\ 0 \end{pmatrix} \qquad (8)$$

The solution is $\lambda_{12} = 0$; $\lambda_{22} = 3.574$ kmol/h; $\lambda_{32} = 0$; $\lambda_{42} = 0$    (9)

Component 3: $H_2SO_4$

$$\begin{pmatrix} 1 & 0 & 0 & 0 \\ -0.02 & 1 & 0 & 0 \\ 0 & 0 & 1 & 0 \\ 0 & 0 & 0 & 1 \end{pmatrix} \cdot \begin{pmatrix} \lambda_{13} \\ \lambda_{23} \\ \lambda_{33} \\ \lambda_{43} \end{pmatrix} = \begin{pmatrix} 11.906 \\ 0 \\ 0 \\ 0 \end{pmatrix} \qquad (10)$$

The solution is
$\lambda_{13} = 11.906$ kmol/h; $\lambda_{23} = 0.238$ kmol/h; $\lambda_{33} = 0$; $\lambda_{43} = 0$    (11)

Component 4: $H_2O$

$$\begin{pmatrix} 1 & 0 & -1 & 0 \\ -0.98 & 1 & 0 & 0 \\ 0 & -0.9804 & 1 & 0 \\ 0 & 0 & -0.5995 & 1 \end{pmatrix} \cdot \begin{pmatrix} \lambda_{14} \\ \lambda_{24} \\ \lambda_{34} \\ \lambda_{44} \end{pmatrix} = \begin{pmatrix} 583.728 \\ 21.551 \\ 0 \\ 17.847 \end{pmatrix} \qquad (12)$$

The solution is
$\lambda_{14} = 15,430.0$; $\lambda_{24} = 15,143.0$; $\lambda_{34} = 14,846.2$; $\lambda_{44} = 8,918.1$    (13)

Component 5: $CaSO_4$

$$\begin{pmatrix} 1 & 0 & 0 & 0 \\ 0 & 1 & 0 & 0 \\ 0 & -0.0192 & 1 & 0 \\ 0 & 0 & -0.3997 & 1 \end{pmatrix} \cdot \begin{pmatrix} \lambda_{15} \\ \lambda_{25} \\ \lambda_{35} \\ \lambda_{45} \end{pmatrix} = \begin{pmatrix} 0 \\ 11.898 \\ 0 \\ 0 \end{pmatrix} \qquad (14)$$

The solution is    $\lambda_{15} = 0$; $\lambda_{25} = 11.898$; $\lambda_{35} = 0.228$; $\lambda_{45} = 0.091$    (15)

Component 6: Free Carbon, C

$$\begin{pmatrix} 1 & 0 & 0 & 0 \\ 0 & 1 & 0 & 0 \\ 0 & -0.0004 & 1 & 0 \\ 0 & 0 & -0.0008 & 1 \end{pmatrix} \cdot \begin{pmatrix} \lambda_{16} \\ \lambda_{26} \\ \lambda_{36} \\ \lambda_{46} \end{pmatrix} = \begin{pmatrix} 0 \\ 0.225 \\ 0 \\ 0 \end{pmatrix} \qquad (16)$$

The solution is    $\lambda_{16} = 0$; $\lambda_{26} = 0.225$; $\lambda_{36} = 0$; $\lambda_{46} = 0$    (17)

Component 7: $CaSO_4.\frac{1}{2}H_2O$

$$\begin{pmatrix} 1 & 0 & 0 & 0 \\ 0 & 1 & 0 & 0 \\ 0 & 0 & 1 & 0 \\ 0 & 0 & 0 & 1 \end{pmatrix} \cdot \begin{pmatrix} \lambda_{17} \\ \lambda_{27} \\ \lambda_{37} \\ \lambda_{47} \end{pmatrix} = \begin{pmatrix} 0 \\ 0 \\ 0 \\ 11.898 \end{pmatrix} \qquad (18)$$

The solution is
$$\lambda_{17} = 0; \ \lambda_{27} = 0; \ \lambda_{37} = 0; \ \lambda_{47} = 11.898 \qquad (19)$$

Component 8: $CO_2$

$$\begin{pmatrix} 1 & 0 & 0 & 0 \\ 0 & 1 & 0 & 0 \\ 0 & 0 & 1 & 0 \\ 0 & 0 & 0 & 1 \end{pmatrix} \cdot \begin{pmatrix} \lambda_{18} \\ \lambda_{28} \\ \lambda_{38} \\ \lambda_{48} \end{pmatrix} = \begin{pmatrix} 0 \\ 3.574 \\ 0 \\ 0 \end{pmatrix} \qquad (20)$$

The solution is

$$\lambda_{18} = 0; \ \lambda_{28} = 3.574; \ \lambda_{38} = 0; \ \lambda_{48} = 0 \qquad (21)$$

**Table 5.14.1: Summary of Component Inlets to Process Units, $\lambda_{ik}$, kmol/h**

| Component | Code | Acid Mixer | Lime Neutraliser | Slurry Separator | Roaster |
|---|---|---|---|---|---|
| $Ca(OH)_2$ | 1 | 0 | 8.234 | 0 | 0 |
| $CaCO_3$ | 2 | 0 | 3.574 | 0 | 0 |
| $H_2SO_4$ | 3 | 11.906 | 0.238 | 0 | 0 |
| $H_2O$ | 4 | 15,430 | 15,143 | 14,846 | 8,918 |
| $CaSO_4$ | 5 | 0 | 11.898 | 0.228 | 0.045 |
| C | 6 | 0 | 0.225 | 0 | 0 |
| $CaSO_4.\frac{1}{2}H_2O$ | 7 | 0 | 0 | 0 | 11.898 |
| $CO_2$ | 8 | 0 | 3.574 | 0 | 0 |

### 5.3: Energy Balances across Process Units

The energy balance, within a specified boundary, is given, generally, as

$Energy\ In + Energy\ Generated$
$$= Energy\ Out + Energy\ Accummulated \qquad (5.72)$$

A typical engineering process can be represented as shown below.

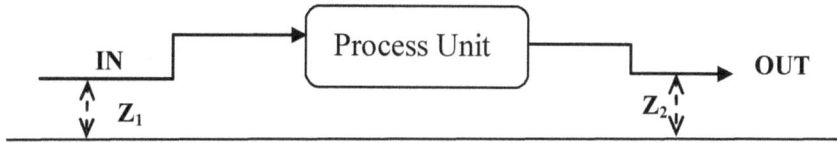

$Z_1$ and $Z_2$ are the elevations of the inlet and outlet streams, in relation to a common reference plane. If the heat absorbed or released in the process unit is designated by Q, while the work done by or on the unit is represented by W, equation (5.72) may be written, on a unit mass basis, for an open, steady state flow system, as

$$U_1 + P_1V_1 + \frac{u_1^2}{2} + gZ_1 + Q = U_2 + P_2V_2 + \frac{u_2^2}{2} + gZ_2 + W \quad (5.73)$$

where U is the internal energy per unit mass, PV is the flow work per unit mass, $\frac{u^2}{2}$ is the kinetic energy per unit mass, gZ is the potential energy per unit mass, and subscripts 1 and 2 denote input and output terminals in the process unit

### 5.3.1: The Thermal Energy Balance

The kinetic and potential energies are, usually, negligible in comparison to other energy terms, in most chemical engineering processes. In such cases, equation (5.73) reduces to

$$U_1 + P_1V_1 + Q = U_2 + P_2V_2 + W \quad (5.74)$$

Since H = U + PV, equation (5.74) can be further reduced to

$$H_1 + Q = H_2 + W \quad (5.75)$$

For processes in which no work is done by or on the process unit, W = 0, so that equation (5.75) becomes

$$Q = H_2 - H_1 \quad (5.76)$$

Equation (5.76) is, also, known as the heat balance equation.

When chemical reaction is involved, the heat absorbed or evolved term, Q, consists of two parts. That is

$$Q = Q_{reaction} + Q_{sensible\ heat} \quad (5.77)$$

$Q_{reaction}$ is the heat absorbed or released as a result of chemical reaction while $Q_{sensible\ heat}$ is the heat absorbed or released otherwise.
The various methods for calculating the heats of reaction and of transition as well as sensible heat have been outlined earlier in the chapter on thermodynamics.

To employ these various heats for an energy balance in a chemical engineering process is to, merely sum them up as they occur in the inputs and outputs and for all components. In terms of rates of flow of mass and energy in a given process stream, this summation gives

$$\sum_i F_i H_i + \sum_i \frac{dQ_i}{dt} + \sum_i \frac{dW_i}{dt} = 0 \qquad (5.78)$$

where F is the mass flow rate, Q the heat lost or gained, W the work done and subscript $i$ denotes component $i$ and $t$, the time variable.

Each enthalpy term, H, can be made up of the sensible enthalpy and the enthalpy of transition.

### 5.3.2: Application of the Thermal Energy Balance to a Process Unit

For a typical process unit, such as the one illustrated below, in which there are material flows, $F_1$, $F_2$, $F_3$ and the associated enthalpies, $H_1$, $H_2$ and $H_3$ with heat absorption or loss of Q, the thermal energy balance equation can be written as

$$F_1 H_1 + F_2 H_2 + Q - F_3 H_3 = 0 \qquad (5.79)$$

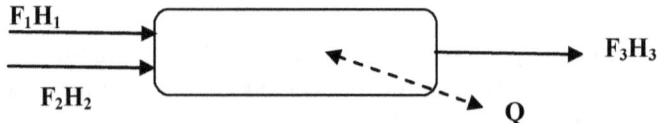

It is usual that the flow rates can be obtained from a material balance across the process unit. For the energy terms, however, one or more of them may be unknown. The unknown energy term may be, either, Q or one or more of the $H_i$. Thus, two types of problems may be encountered in a thermal energy balance across a process unit.

1. If the thermal conditions of the process streams are known, the problem becomes that of determining the heat released or absorbed, Q.
2. If Q is known but one of the inlet or outlet thermal conditions is unknown, this unknown will have to be determined.

Both of these types of problems may be solved as follows.

### 5.3.3: Thermal Energy Balance by the Method of Constraints

The procedure for the thermal energy balance, in this method, is the same as for the material balance except that the energy balance equations are now added to those of the material balance.

## 1. Equations and Constraints

- Material Balance Equations (as before)
- Mole Fraction Constraints (as before)
- Equipment Constraints (as before)
- Energy Balance Constraints $(Q - F_3H_3 + F_2H_2 + F_1H_1 = 0)$
- Enthalpy Equations $[H_1 = f_1 (T_1, P_1, x_1); H_2 = f_2 (T_2, P_2, x_2); H_3 = f_3(T_3, P_3, x_3)]$
- Number of Variables (as before)
- Design/Decision Variables (as before)
- Solution of Equations.(as before)

**Illustrative Example 5.15**

Determine the energy balance for the problem of Example 5.12 assuming that the LPG and air enter the burner at 30 C and 1 atm pressure and flue gases leave at 500 C.

**Answer**            .

The burner was represented schematically as follows:

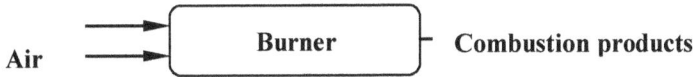

The equation to be solved, in addition to the mass balance equations, is
$$F_1H_1 + F_2H_2 + Q - F_3H_3 = 0 \qquad (1)$$
where $H_1 = f_1 (T_1, P_1, x_1); H_2 = f_2 (T_2, P_2, x_2); H_3 = f_3(T_3, P_3, x_3)$

The enthalpy, $H_i$, for any stream, is given by
$$H_i = \sum_{i=1}^{N} x_{ij} \left( H_j^o + \int_{T_o}^{T} Cp_j dT \right) \qquad (2)$$

We shall assume a reference temperature of 25 C or 298 K. The other process temperatures are
$$T_1 = 273 + 30 = 303 \ K; \quad T_2 = 273 + 30 = 303 \ K; \quad T_3 = 273 + 500$$
$$= 773 \ K \qquad (3)$$

In order to evaluate equations (1) and (2) above, we shall require data on mole fractions, specific heat capacities and heats of formation of every component involved in the combustion. The composition and heats of formation of the LPG, air and flue gases, obtained from Example 5.12, are shown below

**Table 5.15.1: Composition and Thermal Properties of Inlet LPG**

| Component | Mole Fraction | Molecular Weight | Heat of Formation, kJ/kmol |
|---|---|---|---|
| Ethane, $C_2H_6$ | 0.00081 | 30 | - 84.74 |
| Propane, $C_3H_8$ | 0.32825 | 44 | - 103.92 |
| Iso-butane, $C_4H_{10}$ | 0.24204 | 58 | - 134.61 |
| n-butane, $C_4H_{10}$ | 0.42291 | 58 | - 126.23 |
| Iso-pentane, $C_5H_{12}$ | 0.00599 | 72 | - 154.58 |
| Total | 1.00000 | | - |

**Table 5.15.2: Composition and Thermal Properties of Inlet Air**

| Component | Mole Fraction | Molecular Weight | Heat of Formation, kJ/kmol |
|---|---|---|---|
| $O_2$ | 0.21 | 32 | 0 |
| $N_2$ | 0.79 | 28 | 0 |
| Total | 1.00 | | |

**Table 5.15.3: Composition and Thermal Properties of Flue Gases**

| Component | Mole Fraction | Molecular Weight | Heat of Formation, kJ/kmol |
|---|---|---|---|
| $O_2$ | 0.00886 | 32 | 0 |
| $CO_2$ | 0.11368 | 44 | - 393.77 |
| $H_2O$ | 0.14457 | 18 | - 242.00 |
| $N_2$ | 0.73290 | 28 | 0 |
| Total | 1.00001 | | |

The specific heat capacities of all the components are listed below

**Table 5.15.4: Specif Heat Capacities of Air Components**

| Component | Specific Heat Capacity, J/mol.K |
|---|---|
| $O_2$ | $Cp = 28.106 - 3.68 \times 10^{-6}T + 1.745 \times 10^{-5}T^2 - 1.065 \times 10^{-8} T^3$ |
| $N_2$ | $Cp = 35.150 - 1.356 \times 10^{-2}T + 2.679 \times 10^{-5}T^2 - 1.168 \times 10^{-8} T^3$ |

## Table 5.15.5:  Specific Heat Capacities of LPG Components

| Component | Specific Heat Capacity, J/mol.K |
|---|---|
| Ethane, $C_2H_6$ | $Cp = 5.409 + 0.1781T + 0.6937x\ 10^{-4}T^2$ $+0.8712\ x\ 10^{-8}\ T^3$ |
| Propane, $C_3H_8$ | $Cp = -4.224 + 0.3062T - 1.586\ x\ 10^{-4}T^2$ $+ 3.214\ x\ 10^{-8}T^3$ |
| Iso-butane, $C_4H_{10}$ | $Cp = -1.390 + 0.3847T - 1.846\ x\ 10^{-4}T^2$ $+ 2.895\ x\ 10^{-8}T^3$ |
| n-butane, $C_4H_{10}$ | $Cp = 9.487 + 0.3313T - 1.108\ x\ 10^{-4}T^2$ $- 0.2821\ x\ 10^{-8}T^3$ |
| Iso-pentane, $C_5H_{12}$ | $Cp = -9.525 + 0.5066T - 2.729\ x\ 10^{-4}T^2$ $+ 5.723\ x\ 10^{-8}T^3$ |

Thus, at 298 K, from equation 2, for $O_2$

$$H_{O_2} = 0.21\left[0 + 28.106(303 - 298) - 3.68\ x\ 10^{-6}\frac{(303^2 - 298^2)}{2}\right]$$

$$+0.21\left[1.745\ x\ 10^{-5}x\frac{(303^3 - 298^3)}{3} - 1.065\ x\ 10^{-8}x\frac{(303^4 - 298^4)}{4}\right]$$

$$= 29.510 + 1.351 = 30.861\frac{J}{mol\ air} \qquad (3)$$

From equation 2, for $N_2$

$$H_{N_2} = 0.79\left[0 + 35.150(303 - 298) - 1.356\ x\ 10^{-2}\frac{(303^2 - 298^2)}{2}\right]$$

$$+0.79\left[2.679\ x\ 10^{-5}x\frac{(303^3 - 298^3)}{3} - 1.168\ x\ 10^{-8}x\frac{(303^4 - 298^4)}{4}\right]$$

$$= 122.747 + 8.304 = 131.051\frac{J}{mol\ air} \qquad (4)$$

Then for the air inlet stream

$$H_{Air} = 30.861 + 131.051 = 161.912\ \frac{J}{mol\ air} \qquad (5)$$

Still at 298 K, from equation 2, for $C_2H_6$

$$H_{C_2H_6} = 0.00081\left[-84,740 + 5.409\ x\ (303 - 298)\right.$$

$$+ 0.1781x\frac{(303^2 - 298^2)}{2}\right]$$

$$+0.00081 \left[ 0.6937 \; x \; 10^{-4}x \cdot \frac{(303^3 - 298^3)}{3} \right.$$

$$\left. + 0.8712 \; x \; 10^{-8}x \cdot \frac{(303^4 - 298^4)}{4} \right]$$

$$= -68.401 + 0.026 = -68.375 \frac{J}{mol \; LPG} \tag{6}$$

From equation 2, for $C_3H_8$

$$H_{C3H8} = 0.32825 \left[ -103{,}920 - 4.224 \; x \; (303 - 298) \right.$$

$$\left. + 0.3062 \; x \cdot \frac{(303^2 - 298^2)}{2} \right]$$

$$-0.32825 \left[ 1.586 \; x \; 10^{-4} \; x \cdot \frac{(303^3 - 298^3)}{3} - 3.214 \; x \; 10^{-8}x \cdot \frac{(303^4 - 298^4)}{4} \right]$$

$$= -33{,}967.656 - 22.074 = -33{,}989.730 \frac{J}{mol \; LPG} \tag{7}$$

From equation 2, for iso-$C_4H_{10}$

$$H_{iso-C4H10} = 0.24204 \left[ -134{,}610 - 1.390 \; x \; (303 - 298) \right.$$

$$\left. + 0.3847x \cdot \frac{(303^2 - 298^2)}{2} \right]$$

$$-0.24204 \left[ 1.846 \; x \; 10^{-4}x \cdot \frac{(303^3 - 298^3)}{3} - 2.895 \; x \; 10^{-8}x \cdot \frac{(303^4 - 298^4)}{4} \right]$$

$$= -32{,}442.785 - 19.223 = -32{,}462.008 \frac{J}{mol \; LPG} \tag{8}$$

From equation 2, for n-$C_4H_{10}$

$$H_{n-C4H10} = 0.42291 \left[ -126{,}230 + 9.487 \; x \; (303 - 298) \right.$$

$$\left. + 0.3313 \; x \cdot \frac{(303^2 - 298^2)}{2} \right]$$

$$-0.42291 \left[ 1.108 \; x \; 10^{-4}x \cdot \frac{(303^3 - 298^3)}{3} + 0.2821 \; x \; 10^{-8}x \cdot \frac{(303^4 - 298^4)}{4} \right]$$

$$= -53{,}153.353 - 21.319 = -53{,}174.672 \frac{J}{mol \; LPG} \tag{9}$$

From equation 2, for iso-$C_5H_{12}$

$$H_{iso-C5H12} = 0.00599\left[-154,580 - 9.525 \; x \; (303 - 298)\right.$$
$$+ 0.5066 \; x \frac{(303^2 - 298^2)}{2}\right]$$
$$-0.00599\left[2.729 \; x \; 10^{-4}x\frac{(303^3 - 298^3)}{3} - 5.723 \; x \; 10^{-8}x\frac{(303^4 - 298^4)}{4}\right]$$
$$= -921.660 - 0.692 = -922.352\frac{J}{mol \; LPG} \qquad (10)$$

Then for the LPG inlet stream

$$H_{LPG} = -68.375 - 33,989.730 - 32,462.008 - 53,174.672 - 922.352$$
$$= -120,617.137 \; \frac{J}{mol \; LPG} \qquad (11)$$

**Table 5.15.6: Specific Heat Capacities of Flue Gas Components**

**Component    Specific Heat Capacity, J/mol.K**

$O_2$        $Cp = 28.106 - 3.68 \; x \; 10^{-6}T + 1.745 \; x \; 10^{-5}T^2$
              $- 1.065 \; x \; 10^{-8} \; T^3$

$CO_2$       $Cp = 19.795 + 7.343 \; x \; 10^{-2}T - 5.601 \; x \; 10^{-5}T^2$
              $+ 1.715 \; x \; 10^{-8} \; T^3$

$H_2O$       $Cp = 32.243 + 1.923 \; x \; 10^{-3}T + 1.055 \; x \; 10^{-5}T^2$
              $- 0.3596 \; x \; 10^{-8} \; T^3$

$N_2$         $Cp = 31.150 - 1.356 \; x \; 10^{-2}T + 2.679 \; x \; 10^{-5}T^2$
              $- 1.168 \; x \; 10^{-8} \; T^3$

At 773 K, from equation 2, for $O_2$

$$H_{O_2} = 0.00886\left[0 + 28.106(773 - 298) - 3.68 \; x \; 10^{-6}\frac{(773^2 - 298^2)}{2}\right]$$
$$+0.00886\left[1.745 \; x \; 10^{-5}x\frac{(773^3 - 298^3)}{3} - 1.065 \; x \; 10^{-8}x\frac{(773^4 - 298^4)}{4}\right]$$
$$= 118.276 + 14.204 = 132.480\frac{j}{mol \; flue \; gas} \qquad (12)$$

From equation 2, for $CO_2$

$$
H_{CO_2} = 0.11368 \left[ -393,770 + 19.795 \, x \, (773 - 298) \right.
$$
$$
\left. + 7.343 \, x \, 10^{-2} \frac{(773^2 - 298^2)}{2} \right]
$$

$$
-0.11368 \left[ 5.601 \, x \, 10^{-5} x \frac{(773^3 - 298^3)}{3} - 1.715 \, x \, 10^{-8} x \frac{(773^4 - 298^4)}{4} \right]
$$

$$
= -41,571.587 - 753.972 = -42,325.559 \frac{J}{mol \ flue \ gas} \tag{13}
$$

From equation 2, for $H_2O$

$$
H_{H_2O} = 0.14457 \left[ -242,000 + 32.243 \, x \, (773 - 298) \right.
$$
$$
\left. + 1.923 \, x \, 10^{-3} \frac{(773^2 - 298^2)}{2} \right]
$$

$$
+0.14457 \left[ 1.055 \, x \, 10^{-5} x \frac{(773^3 - 298^3)}{3} - 0.3596 \, x \, 10^{-8} x \frac{(773^4 - 298^4)}{4} \right]
$$

$$
= -32,701.074 - 175.994 = -32,877.068 \frac{J}{mol \ flue \ gas} \tag{14}
$$

From equation 2, for $N_2$

$$
H_{N_2} = 0.73290 \left[ 0 + 35.150( \, 773 - 298) - 1.356 \, x \, 10^{-2} \frac{(773^2 - 298^2)}{2} \right]
$$

$$
+0.73290 \left[ 2.679 \, x \, 10^{-5} x \frac{(773^3 - 298^3)}{3} - 1.168 \, x \, 10^{-8} x \frac{(773^4 - 298^4)}{4} \right]
$$

$$
= 9708.796 + 2102.562 = 11,811.358 \frac{J}{mol \ flue \ gas} \tag{15}
$$

Thus for the flue gas stream

$$
H_{flue \ gas} = 132.480 - 42,325.559 - 32,877.068 + 11,811.358
$$
$$
= -63,258.789 \frac{J}{mol \ flue \ gas} \tag{16}
$$

These may be summarized in the Tables below

## Table 5.15.7:  Stream Mole Fractions

| Component | Air Stream, 30 kmol/h | LPG Stream, 1 kmol/h | Flue Gas Stream, 32.338 kmol/h |
|---|---|---|---|
| $O_2$ | 0.21 | | 0.00886 |
| $CO_2$ | | | 0.11368 |
| $H_2O$ | | | 0.14457 |
| $N_2$ | 0.79 | | 0.73290 |
| Ethane, $C_2H_6$ | | 0.00081 | |
| Propane, $C_3H_8$ | | 0.32825 | |
| Iso-butane, $C_4H_{10}$ | | 0.24204 | |
| n-butane, $C_4H_{10}$ | | 0.42291 | |
| Iso-pentane, $C_5H_{12}$ | | 0.00599 | |
| Totals | 1.00 | 1.00000 | 1.0001 |

Thus, from equations (1), (5), (11) and (16)

$$F_{air}H_{air} + F_{LPG}H_{LPG} + Q - F_{flue\ gas}H_{flue\ gas} = 0 \qquad (17)$$
$$Q = F_{flue\ gas}H_{flue\ gas} - F_{air}H_{air} - F_{LPG}H_{LPG}$$
$$= 32.338\ x - 63{,}258.789 - 30\ x\ 161.912 - 1\ x - 120{,}617.137$$
$$= -1{,}929{,}902.942k\ J/h \qquad Heat\ energy\ released \qquad Answer$$

## Table 5.15.8:  Component Enthalpies, kJ/kmol

| Component | Air Stream, 298 K | LPG Stream, 298 K | Flue Gas Stream, 773 K |
|---|---|---|---|
| $O_2$ | 30.861 | | 132.480 |
| $CO_2$ | | | -42,325.559 |
| $H_2O$ | | | -32,877.008 |
| $N_2$ | 131.051 | | 11,811.358 |
| Ethane, $C_2H_6$ | | -68.375 | |
| Propane, $C_3H_8$ | | -33,989.730 | |
| Iso-butane, $C_4H_{10}$ | | -32,462.008 | |
| n-butane, $C_4H_{10}$ | | -53,174.672 | |
| Iso-pentane, $C_5H_{12}$ | | -922.352 | |
| Totals | 161.912 | -120,617.137 | -63,258.729 |

### 5.3.4: *Thermal Energy Balance by the Method of Split Fraction Coefficients*

Although the thermal energy balance can be obtained using standard procedures once the mass balance is known, the Split Fraction Coefficient method may, also, be used. The procedure for the thermal energy balance, in this method, follows closely that for the material balance, discussed in Section 5.2.3. It involves

1.  developing the applicable information flow diagram
2.  developing the energy balance equations in terms of the split fraction coefficients and fresh feeds
3.  estimating the energy per unit mass of component $k$ in fresh feeds and in streams entering and leaving any process unit $j$
4.  estimating the total energy of fresh feeds and of streams entering and leaving any process unit $j$
5.  solving the resulting simultaneous equations using the values of enthalpies and the material balance of the components.

### Illustrative Example 5.16

Recalculate the energy balance for the problem of Example 5.15 using the Split Fraction Coefficient method. The LPG and air enter the burner at 30 C and 1 atm pressure and flue gases leave at 500 C.

**Answer**

The material balance for the burner may be represented schematically as follows:

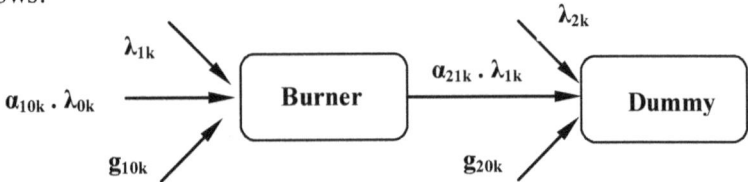

The material balance is given as

$$\lambda_{1k} = g_{10k} + \alpha_{10k} \cdot \lambda_{0k} \tag{1}$$

$$\lambda_{2k} = g_{20k} + \alpha_{21k} \cdot \lambda_{1k} \tag{2}$$

The resulting matrix is

$$\begin{pmatrix} -\alpha_{10k} & 1 & 0 \\ 0 & -\alpha_{21k} & 1 \end{pmatrix} \cdot \begin{pmatrix} \lambda_{0k} \\ \lambda_{1k} \\ \lambda_{2k} \end{pmatrix} = \begin{pmatrix} g_{10k} \\ g_{20k} \end{pmatrix} \tag{3}$$

The estimated split fraction coefficients and fresh feeds are listed in Table 5.16.1. The $\alpha_{21k}.\lambda_{1k}$ have already been calculated from the previous example so that calculating $\alpha_{21k}$ is unnecessary. Rather than solve the matrices of equation (3) for each of the nine components, Table 5.16.1 shows a much simpler solution of equations (1) and (2).

**Table 5.16.1: Split Fraction Coefficients and Fresh Feeds for the Material Balance according to Equation (3)**
**Basis: 1kmol/h LPG, 30 kmol/h Air**

| Component | Code, k | $g_{10k}$ | $\lambda_{0k}$ | $\alpha_{10k}$ | $\lambda_{1k}$ | $g_{20k}$ | $\alpha_{21k}.\lambda_{1k}$ | $\lambda_{2k}$ |
|---|---|---|---|---|---|---|---|---|
| $C_2H_6$ | 1 | 0.00081 | 0 | 0 | 0.00081 | 0 | 0 | 0 |
| $C_3H_8$ | 2 | 0.32825 | 0 | 0 | 0.32825 | 0 | 0 | 0 |
| $C_4H_{10}$ | 3 | 0.24204 | 0 | 0 | 0.24204 | 0 | 0 | 0 |
| $C_4H_{10}$ | 4 | 0.42291 | 0 | 0 | 0.42291 | 0 | 0 | 0 |
| $C_5H_{12}$ | 5 | 0.00599 | 0 | 0 | 0.00599 | 0 | 0 | 0 |
| $O_2$ | 6 | 6.3 | 0 | 0 | 6.3 | 0 | 0.28651 | 0.28651 |
| $N_2$ | 7 | 23.7 | 0 | 0 | 23.7 | 0 | 23.7005 | 23.7005 |
| $CO_2$ | 8 | | 0 | 0 | | 0 | 3.67618 | 3.67618 |
| $H_2O$ | 9 | | 0 | 0 | | 0 | 4.67510 | 4.67510 |
| Totals | | 31 | | | 31 | | 32.33831 | 32.33831 |

The information flow diagram for the energy balance is

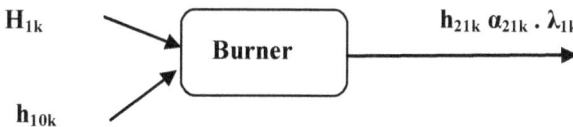

Both $h_{iok}$ and $h_{jik}$ are, each, given by

$$h_{ik}, \; h_{iok} \; or \; h_{jik} = h_{iok}^o \left( or \; h_{jik}^o \right) + \int_{T_o}^{T} Cp_k dT \tag{4}$$

where $h_{iok}^o$ (or $h_{jik}^o$) is the specific enthalpy of component $k$ at the reference temperature $T_o$ and $h_{iok}$ (or $h_{jik}$) is the specific enthalpy of component $k$ in relation to process unit $i$ at temperature T. We may, also, in similarity to the material balance procedures, define

$$h_{jik} = \beta_{jik}.H_{ik} \tag{5}$$

The energy balance is, therefore,

$$H_{ik} = h_{ik} . \lambda_{ik} = h_{i0k}.g_{i0k} + h_{jik}\alpha_{jik}\lambda_{ik} \tag{6}$$

## Estimating the Specific Enthalpies of the Components

The specific enthalpies at 298 are calculated for $O_2$, from equation (4), as

$$h_{O_2} = \left[0 + 28.106(303 - 298) - 3.68 \, x \, 10^{-6}\frac{(303^2 - 298^2)}{2}\right]$$

$$+ \left[1.745 \, x \, 10^{-5}x\frac{(303^3 - 298^3)}{3} - 1.065 \, x \, 10^{-8}x\frac{(303^4 - 298^4)}{4}\right]$$

$$= 140.524 + 6.433 = 146.957\frac{J}{mol \, air} \tag{7}$$

For $N_2$

$$h_{N_2} = \left[0 + 35.150(303 - 298) - 1.356 \, x \, 10^{-2}\frac{(303^2 - 298^2)}{2}\right]$$

$$+ \left[2.679 \, x \, 10^{-5}x\frac{(303^3 - 298^3)}{3} - 1.168 \, x \, 10^{-8}x\frac{(303^4 - 298^4)}{4}\right]$$

$$= 155.376 + 10.511 = 165.887\frac{J}{mol \, air} \tag{8}$$

Still at 298 K, from equation (4), for $C_2H_6$

$$h_{C_2H_6} = \left[-84,740 + 5.409 \, x \, (303 - 298) + 0.1781x\frac{(303^2 - 298^2)}{2}\right]$$

$$+ \left[0.6937 \, x \, 10^{-4}x\frac{(303^3 - 298^3)}{3} + 0.8712 \, x \, 10^{-8}x\frac{(303^4 - 298^4)}{4}\right]$$

$$= -84,445.679 + 32.099 = -84,413.580\frac{J}{mol \, LPG} \tag{9}$$

For $C_3H_8$

$$h_{C3H8} = \left[-103,920 - 4.224 \, x \, (303 - 298) + 0.3062 \, x\frac{(303^2 - 298^2)}{2}\right]$$

$$- \left[1.586 \, x \, 10^{-4} \, x\frac{(303^3 - 298^3)}{3} - 3.214 \, x \, 10^{-8}x\frac{(303^4 - 298^4)}{4}\right]$$

$$= -103,481.054 - 67.248 = -103,548.302\frac{J}{mol \, LPG} \tag{10}$$

For iso-$C_4H_{10}$

$$h_{iso-C4H10} = \left[ -134{,}610 - 1.390 \, x \, (303 - 298) \right.$$

$$\left. + 0.3847 x \frac{(303^2 - 298^2)}{2} \right]$$

$$- \left[ 1.846 \, x \, 10^{-4} x \frac{(303^3 - 298^3)}{3} - 2.895 \, x \, 10^{-8} x \frac{(303^4 - 298^4)}{4} \right]$$

$$= -134{,}038.940 - 79.421 = -134{,}118.361 \frac{J}{mol \, LPG} \qquad (11)$$

For n-$C_4H_{10}$

$$h_{n-C4H10} = \left[ -126{,}230 + 9.487 \, x \, (303 - 298) \right.$$

$$\left. + 0.3313 \, x \frac{(303^2 - 298^2)}{2} \right]$$

$$- \left[ 1.108 \, x \, 10^{-4} x \frac{(303^3 - 298^3)}{3} + 0.2821 \, x \, 10^{-8} x \frac{(303^4 - 298^4)}{4} \right]$$

$$= -125{,}684.786 - 50.410 = -125{,}735.196 \frac{J}{mol \, LPG} \qquad (12)$$

For iso-$C_5H_{12}$

$$h_{iso-C5H12} = \left[ -154{,}580 - 9.525 \, x \, (303 - 298) \right.$$

$$\left. + 0.5066 \, x \frac{(303^2 - 298^2)}{2} \right]$$

$$- \left[ 2.729 \, x \, 10^{-4} x \frac{(303^3 - 298^3)}{3} - 5.723 \, x \, 10^{-8} x \frac{(303^4 - 298^4)}{4} \right]$$

$$= -153{,}866.459 - 115.452 = -153{,}981.970 \frac{J}{mol \, LPG} \qquad (13)$$

Using the same specific heat capacity equations of Table 5.15.6 for the flue gas components, the specific enthalpies are, also, obtained as follows

At 773 K, from equation (4), for $O_2$

$$h_{O_2} = \left[ 0 + 28.106(773 - 298) - 3.68 \, x \, 10^{-6} \frac{(773^2 - 298^2)}{2} \right.$$

273

$$+ \left[1.745 \times 10^{-5}x\frac{(773^3 - 298^3)}{3} - 1.065 \times 10^{-8}x\frac{(773^4 - 298^4)}{4}\right]$$

$$= 13{,}349.436 + 1603.160 = 14{,}952.596\frac{j}{mol\ flue\ gas} \qquad (14)$$

For $CO_2$

$$h_{CO_2} = \left[-393{,}770 + 19.795 \times (773 - 298) + 7.343 \times 10^{-2}\frac{(773^2 - 298^2)}{2}\right]$$

$$- \left[5.601 \times 10^{-5}x\frac{(773^3 - 298^3)}{3} - 1.715 \times 10^{-8}x\frac{(773^4 - 298^4)}{4}\right]$$

$$= -365{,}689.541 - 6{,}632.407 = -372{,}321.948\ \frac{J}{mol\ flue\ gas} \qquad (15)$$

For $H_2O$

$$h_{H_2O} = \left[-242{,}000 + 32.243 \times (773 - 298)\right.$$

$$\left. + 1.923 \times 10^{-3}\frac{(773^2 - 298^2)}{2}\right]$$

$$+ \left[1.055 \times 10^{-5}x\frac{(773^3 - 298^3)}{3} - 0.3596 \times 10^{-8}x\frac{(773^4 - 298^4)}{4}\right]$$

$$= -226{,}195.435 - 1217.362 = -227{,}412.797\frac{J}{mol\ flue\ gas} \qquad (16)$$

For $N_2$

$$h_{N_2} = \left[0 + 35.150(773 - 298) - 1.356 \times 10^{-2}\frac{(773^2 - 298^2)}{2}\right.$$

$$+ \left[2.679 \times 10^{-5}x\frac{(773^3 - 298^3)}{3} - 1.168 \times 10^{-8}x\frac{(773^4 - 298^4)}{4}\right]$$

$$= 13{,}247.095 + 2{,}868.825 = 16{,}115.920\frac{J}{mol\ flue\ gas} \qquad (17)$$

$$h_{C_2H_6} = \left[-84{,}740 + 5.409 \times (773 - 298) + 0.1781x\frac{(773^2 - 298^2)}{2}\right.$$

$$+ \left[0.6937 \times 10^{-4}x\frac{(773^3 - 298^3)}{3} + 0.8712 \times 10^{-8}x\frac{(773^4 - 298^4)}{4}\right]$$

$$= -36{,}868.764 + 10{,}828.967 = -26{,}039.797\frac{J}{mol\ LPG} \qquad (18)$$

For $C_3H_8$

$$h_{C3H8} = \left[ -103{,}920 - 4.224 \; x \; (773 - 298) + 0.3062 \; x \frac{(773^2 - 298^2)}{2} \right]$$

$$- \left[ 1.586 \; x \; 10^{-4} \; x \frac{(773^3 - 298^3)}{3} - 3.214 \; x \; 10^{-8} x \frac{(773^4 - 298^4)}{4} \right]$$

$$= -28{,}040.603 - 20{,}214.080 = -48{,}254.683 \frac{J}{mol \; LPG} \qquad (19)$$

For iso-$C_4H_{10}$

$$h_{iso-C4H10} = \left[ -134{,}610 - 1.390 \; x \; (773 - 298) \right.$$

$$\left. + 0.3847x \frac{(773^2 - 298^2)}{2} \right]$$

$$- \left[ 1.846 \; x \; 10^{-4}x \frac{(773^3 - 298^3)}{3} - 2.895 \; x \; 10^{-8}x \frac{(773^4 - 298^4)}{4} \right]$$

$$= -37{,}416.996 - 24{,}266.226 = -61{,}683.222 \frac{J}{mol \; LPG} \qquad (20)$$

For n-$C_4H_{10}$

$$h_{n-C4H10} = \left[ -126{,}230 + 9.487 \; x \; (773 - 298) \right.$$

$$\left. + 0.3313 \; x \frac{(773^2 - 298^2)}{2} \right]$$

$$- \left[ 1.108 \; x \; 10^{-4}x \frac{(773^3 - 298^3)}{3} + 0.2821 \; x \; 10^{-8}x \frac{(773^4 - 298^4)}{4} \right]$$

$$= -37{,}453.379 - 16{,}327.987 = -53{,}781.366 \frac{J}{mol \; LPG} \qquad (21)$$

For iso-$C_5H_{12}$

$$h_{iso-C5H12} = \left[ -154{,}580 - 9.525 \; x \; (773 - 298) \right.$$

$$\left. + 0.5066 \; x \frac{(773^2 - 298^2)}{2} \right]$$

$$- \left[ 2.729 \; x \; 10^{-4}x \frac{(773^3 - 298^3)}{3} - 5.723 \; x \; 10^{-8}x \frac{(773^4 - 298^4)}{4} \right]$$

$$= -30{,}244.333 - 34{,}613.750 = -64{,}858.083 \frac{J}{mol \; LPG} \qquad (22)$$

These are summarized in Table 5.16.2.

### Table 5.16.2: Component Specific Enthalpies, $h_{jik}$, kJ/kmol

| Component | Air Stream, $h_{10k}$, 298 K | LPG Stream, $h_{10k}$, 298 K | Flue Gas Stream, $h_{21k}$, 773 K |
|---|---|---|---|
| $O_2$ | 146.957 | | 14,952.596 |
| $N_2$ | 165.887 | | 16,115.920 |
| $CO_2$ | | | −372,321.948 |
| $H_2O$ | | | −227,412.797 |
| Ethane, $C_2H_6$ | | −84,413.580 | −26,039.797 |
| Propane, $C_3H_8$ | | −103,548.302 | −48,254.683 |
| Iso-butane, $C_4H_{10}$ | | −134,118.361 | −61,683.222 |
| n-butane, $C_4H_{10}$ | | −125,735.196 | −53,781.366 |
| Iso-pentane, $C_5H_{12}$ | | −153,981.970 | −64,858.083 |

The energy balance, according to equation (3), may be similarly
tabulated as shown in Tables 5.16.3 and 5.16.4 below

### Table 5.16.3: The Energy Balance of Streams Entering the Burner at 298 K according to Equation (6)
### Basis: 1kmol/h LPG, 30 kmol/h Air

| Component | Code, k | $h_{10k}$, kJ/kmol | $g_{10k} = \lambda_{-1k}$ kmol/h | $H_{1k} = h_{10k} \times g_{10k}$, kJ/h |
|---|---|---|---|---|
| $C_2H_6$ | 1 | −84,413.580 | 0.00081 | - 68.375 |
| $C_3H_8$ | 2 | −103,548.302 | 0.32825 | - 33,989.730 |
| iso-$C_4H_{10}$ | 3 | −134,118.361 | 0.24204 | - 32,462.008 |
| n-$C_4H_{10}$ | 4 | −125,735.196 | 0.42291 | - 53,174.672 |
| Iso-$C_5H_{12}$ | 5 | −153,981.970 | 0.00599 | - 922.352 |
| $O_2$ | 6 | 146.957 | 6.3 | 925.829 |
| $N_2$ | 7 | 165.887 | 23.7 | 3,931.522 |
| $CO_2$ | 8 | | | |
| $H_2O$ | 9 | | | |
| Totals | | | 31 | - 115,759.786 |

**Table 5.16.4: The Energy Balance of Streams Leaving the Burner and Entering the Dummy Unit at 773 K according to Equation (6)**
**Basis: 1kmol/h LPG, 30 kmol/h Air**

| Component | Code, $k$ | $h_{21k}$ | $\alpha_{21k} \cdot \lambda_{1k}$ | $g_{20k}$ | $H_{2k} = h_{20k} \times g_{20k} + h_{21k} \alpha_{21k} \cdot \lambda_{1k}$ |
|---|---|---|---|---|---|
| $C_2H_6$ | 1 | $-26{,}039.797$ | 0 | 0 | 0 |
| $C_3H_8$ | 2 | $-48{,}254.683$ | 0 | 0 | 0 |
| iso-$C_4H_{10}$ | 3 | $-61{,}683.222$ | 0 | 0 | 0 |
| n-$C_4H_{10}$ | 4 | $-53{,}781.366$ | 0 | 0 | 0 |
| Iso-$C_5H_{12}$ | 5 | $-64{,}858.083$ | 0 | 0 | 0 |
| $O_2$ | 6 | $14{,}952.596$ | 0.28651 | 0 | 4,284.068 |
| $N_2$ | 7 | $16{,}115.920$ | 23.70052 | 0 | 381,933.880 |
| $CO_2$ | 8 | $-372{,}321.948$ | 3.67618 | 0 | -1,368,722.499 |
| $H_2O$ | 9 | $-227{,}412.797$ | 4.67510 | 0 | -1,063,173.841 |
| Totals | | | 32.33831 | | -2,045,678.392 |

The energy change on combustion is then

$$Energy\ Leaving - Energy\ entering$$
$$= -2{,}045{,}678.392 - (-115{,}759.786)$$
$$= -1{,}929{,}918.606 \ kJ/h \ Ans.$$

Note that the difference between the results from the Split Fraction Coefficient method and that from the Method of Contraints is only 0.0008 % which is negligible.

## *References for Chapter Five*

1. Green D. W., (Editor); *Perry's Chemical Engineering Handbook*, 6th Edition, Chapter 3; McGraw-Hill Book Compnay, NY., USA (1984)

2. Henley & Rosen (1969) in Sinnott R. K; *An Introduction to Chemical Engineering Design; Chemical Engineering, Vol 6 (SI Units)* by J. M. Coulson & J. F. Richardson; Pergamon Press, Oxford, UK (1983)

3. Hougen O. A., Watson K. M., Ragatz R. A., *Chemical Process Principles Part 1: Material and Energy Balances*; 2nd Edition; Chapters 3, 4 and 5; John Wiley & Sons Inc., NY, USA (1954)

4. Myers A. I., Seider W. D., *Introduction to Chemical Engineering and Computer Calculations*; Chapters 6, 7, 8, 9 and 10; Prentice Hall, N.J., USA (1976)

5. Okori (1976) in Ajibade A. M., *Design Algorithm for a Gas Burner Heated Air Heat Exchanger*; HND Thesis (1989); Dept of Chemical Engineering, I. M. T., Enugu

6. Schmidt A. X., List H. L., *Material and Energy Balances*; Chapter 3; Prentice Hall, N. J., USA, (1962)

7. Sinnott R. K; *An Introduction to Chemical Engineering Design; Chemical Engineering, Vol 6 (SI Units)* by J. M. Coulson & J. F. Richardson; Pergamon Press, Oxford, UK (1983)

8. Treybal R. E; *Mass Transfer Operations*, 3rd Edition; Chapter 7; McGraw-Hill Book Company, NY, USA (1980)

278

# CHAPTER SIX
# THE PROCESS DESIGN REPORT

In Chapter One, it was stated that *"Most products in the market place arose from ideas that were pursued to fruition. Everybody is full of ideas throughout one's life but only those ideas which come with some conception of a process and a strong desire for their implementation will lead to product."*

The previous chapters dealt with the various methods by means of which ideas in chemical processing may be analysed, costed and evaluated.

The results of these analyses, costing and evaluation must be presented as a report, in acceptable form, to third parties who may be financial investors, in-house organisation managers or even governments or regulatory agencies.

The process design report may be part of a feasibility study or it may be the implementation of a decision already taken for an investment to be made. In all cases, it is either a preliminary or a component report to the final design report of the chemical or processing plant in which greater detail and specifications are requitred.

In this chapter, we shall summarise some of the essential features that should be included in a process design report.

The first requirement is to determine what purpose the process design report is aimed to serve. If it is part of a feasibility report, it must be tailored to meet the needs of the feasibility report.

Feasibility reports aim to objectively and rationally uncover the strengths and weaknesses of an existing business or a proposed venture, opportunities and threats, as presented by the environment, the resources required to carry through, and ultimately the prospects for success (Justis, R. T. & Kreigsmann, B., 1979 and Georgakellos, D. A. & Marcis, A. M., 2009).

In its simplest terms, the two criteria which judge the feasibility of a project are *cost* and *value* (ethical and economic) to be attained (Young, G. I. M., 1970). Generally, feasibility studies precede technical development and project implementation.

Of the five factors, abbreviated as TELOS in project management, which determine whether a project should run or not (ibid), namely,

1. T - Technical - Is the project technically possible.
2. E - Economic - Can the project be afforded? Will it increase profit?
3. L - Legal - Is the project legal?
4. O - Organisational - Will the organisation accept the change?
5. S - Scheduling - Can the project be done in time?

process design is, mainly, concerned with the first two. Thus, the process design report must present the technical and economic components of the project as accurately as possible.

If the process design report is a stage before, or a component of, the final equipment and or plant design report, greater emphasis is laid on the technical components of the design without compromising the economic aspects of the project.

The actual details of any process design report will, undoubtably, vary depending on the chemical process and products. All process design reports should, however, contain the essential headings listed in Table 6.1

**Table 6.1: Essential Components of a Process Design Report (Adapted from Humbird et al, 2011).**

| S/No | Item |
|------|------|
| 1 | Executive Summary |
| 2 | Table of Contents <br> • List of Figures <br> • List of Tables |
| 3 | Introduction |
| 4 | Design Basis and Conventions |
| 5 | Process Design and Cost Estimation Methods |
| 6 | Process Economics |
| 7 | Analysis and Discussion |

| 8  | Conclusion |
|----|------------|
| 9  | References |
| 10 | Appendix 1: Individual Equipment Summaries |
| 11 | Appendix 2: Worksheet of the Discounted Cash Flow Rate of Return (DCFRR) and other Profitability Criteria |
| 12 | Appendix 3: Definitions of Process Parameters and Operating Conditions |
| 13 | Appendix 4: Properties of Materials |
| 14 | Appendix 5: Process Flow Diagrams |

### References for Chapter Six

1. Justis, R. T. & Kreigsmann, B. (1979). The feasibility study as a tool for venture analysis. *Business Journal of Small Business Management* 17 (1) 35-42.

2. Georgakellos, D. A. & Marcis, A. M. (2009). Application of the semantic learning approach in the feasibility studies preparation training process. *Information Systems Management*

3. Young, G. I. M. (1970). Feasibility studies. *Appraisal Journal* 38 (3) 376-383.

4. Bentley, L & Whitten, J (2007). *System Analysis & Design for the Global Enterprise*. 7th ed. (p. 417).

5. Michele Berrie (September 2008), *Initiating Phase - Feasibility Study Request and Report*
    26 (3) 231-24
    References 1 to 5 as in Wikipedia, the Free Encyclopaedia, last modified on 10 October 2011 at 09:40.

6. Humbird D., Davis R., Tao L., Kinchin C., Aden A., Schoen P., Lukas J., Olthof B., Worley M., Sexton D., Dudgeon D.; *Process Design and Economics for Biochemical Conversion of Lignocellulosic Biomass to Ethanol – Dilute Acid Pretreatment and Enzymatic Hydrolysis of Corn Stover*; Technical Report NREL/TP-5100-47764; May 2011

## *APPENDIX I: Densities of Various Materials*
## *(www. EngineeringToolBox.com)*

| Material | $(kg/m^3)$ | $(kg/m^3)$ |
|---|---|---|
| ABS resin, pellet | 721 | |
| Acetic acid, liquid | | |
| Acetone | 785 | |
| Acrylic resin | | |
| Adipic acid, powder | 721 | |
| Alcohol, ethyl | | |
| Alcohol, methyl | 785 | |
| Alfalfa, ground | | |
| Almonds, shelled | 481 | 561 |
| Alum powder | | |
| Alumina | 961 | |
| Aluminum hydrate | | |
| Aluminum oxide | 961 | 1602 |
| Aluminum silicate | | 721 |
| Aluminum, powder | 721 | 1281 |
| Aluminum, shavings | | 240 |
| Ammonium nitrate, prill | 721 | 961 |
| Ammunium sulphate | | 929 |
| Apple seed | 513 | |
| Asbestos fibers | | 400 |
| Ash, coal, damp | 721 | 801 |
| Ash, coal, dry | | 721 |
| Asphalt, liquid | 1041 | |
| Aviation fuel (jp-4) | | |
| Bakalite, powder | 481 | 641 |
| Baking powder | | 721 |

283

| | | |
|---|---|---|
| Baking soda | 1121 | 1281 |
| Ball clay | | |
| Bagasse - exiting the final mill | 120 | |
| Bagasse - stacked to 2 metre height (moisture = 44%) | 176 | |
| Bark, wood refuse | 160 | 320 |
| Barley, flour | | 481 |
| Barley, ground | 400 | 481 |
| Barley, kernal | | 641 |
| Barley, malted | 497 | |
| Bauxite, crushed | | 1362 |
| Beans, caster | 577 | |
| Beans, coffee | | 641 |
| Beans, lima | 721 | |
| Beans, navy | | |
| Beans, soy | 721 | 753 |
| Bentonite, lump | | 641 |
| Bentonite, powder | 801 | 961 |
| Bicarbonate of soda | | |
| Blood, dry | 561 | 721 |
| Bone meal | | 961 |
| Borate of lime | 801 | 1121 |
| Borax | | 1121 |
| Boric acid powder | 881 | |
| Bran, oat | | |
| Bran, wheat | 240 | 320 |
| Brewers grain | | |
| Brewers grits | 529 | |
| Bronze chips | | 801 |
| Buckwheat | 545 | 673 |
| Buckwheat flour | | |
| Butter | 865 | |

| | | |
|---|---|---|
| Buttermilk powder | | 481 |
| Cake mix | 481 | 641 |
| Calcium carbide | | |
| Calcium carbonate | 1201 | |
| Calcium oxide | | |
| Cane - whole stick, tangled and tamped down as in a cane transport vehicle | 200 | |
| Cane - whole stick, neatly bundled | 400 | |
| Cane - billetted | 352 | |
| Cane - whole stick tangled, but loosely tipped into cane carrier | 160 | |
| Cane - knifed | 288 | |
| Cane - shredded | 320 | |
| Carbide powder | 1602 | |
| Carbon black powder | | 400 |
| Carbon black, pellet | 320 | 721 |
| Carbon tetrachloride | | |
| Carbon, granulated, activated | 801 | 961 |
| Carbon, graphite | | |
| Casein powder | 561 | 641 |
| Cashew nuts | | 593 |
| Caster beans | 577 | |
| Cat food | | 400 |
| Cellophane, flocking | 80 | |
| Cellulose acetate | | |
| Cellulose, flocking | 24 | 48 |
| Cement powder, portland | | 1522 |
| Cement, clinker | 1201 | 1442 |
| Cereal flake | | |
| Chalk, fine | 1121 | 1201 |
| Chalk, lump | | 1442 |
| Charcoal | 240 | 481 |

| | | |
|---|---|---|
| Chromium ore | | |
| Cinders, coal | 641 | 801 |
| Citric acid | | |
| Clay, attapulgus | 881 | |
| Clay, ball | | |
| Clay, bentonite | 817 | |
| Clay, calcined | | |
| Clay, dicalite | 320 | 801 |
| Clay, kaoline | | 961 |
| Clay, sno-brite | 240 | 801 |
| Clay, whitex | | 801 |
| Clinker, cement | 1281 | |
| Clinker, coal | | 1442 |
| Coal, ground | 641 | |
| Coal, lump | | 881 |
| Coconut, shredded | 320 | 352 |
| Coffee bean, green | | 721 |
| Coffee bean, roasted | 352 | 481 |
| Coffee, ground | | |
| Coke, calcined, petrol | 561 | 721 |
| Copper ore | | |
| Copper oxide | 3043 | |
| Cork, ground | | 240 |
| Corn bran | 208 | |
| Corn cob, ground | | |
| Corn, cracked | 561 | 641 |
| Corn, flaked | | |
| Corn, gern | 336 | |
| Corn, gluten | | 529 |
| Corn, grits | 641 | 721 |

| | | |
|---|---|---|
| Corn, ground | | 561 |
| Corn, meal | 513 | 641 |
| Corn, starch | | 561 |
| Corn, sugar, liquid | 1410 | |
| Corn, sugar, powder | | |
| Corn, whole kernel | 721 | |
| Cotton blossoms | | 400 |
| Cottonseed | 352 | 641 |
| Cottonseed hulls | | |
| Cottonseed meats | 641 | |
| Cottonseed oil | | |
| Cottonseed, meal | 561 | 641 |
| Cream powder | | |
| Cullett, glass | 1922 | |
| Dextrin | | 881 |
| Dextrose | 497 | |
| Diatomacaous earth | | 224 |
| Dicalcium phosphate | 689 | |
| Diesel fuel | | |
| Dirt, dry | 1041 | 1281 |
| Distillars grain | | |
| Dog food, IAMS minichunk | 416 | |
| Dolomite, lump | | 1586 |
| Dolomite, powdered | 721 | |
| Down, goose | | |
| Ebonite, crushed | 1041 | 1121 |
| Emery, crushed | | |
| Epsom salt | 641 | 801 |
| Ethanol | | |
| Ethyl ether | 705 | |

| | | |
|---|---|---|
| Ethylene glycol | | |
| Expancel microsphere | 13 | |
| Farina | | |
| Feathers, goose | 16 | |
| Feed pellets, animal | | 609 |
| Feldspar, ground | 1041 | 1121 |
| Ferrous sulphate | | 1201 |
| Fertilizer, phosphate | 961 | |
| Fish meal | | 641 |
| Flax seed | 641 | 721 |
| Flour, barley | | 481 |
| Flour, corn | 481 | 545 |
| Flour, patent | | |
| Flour, wheat | 481 | 561 |
| Flourospar | | |
| Fluff, poly-fim floc | 24 | 32 |
| Fly ash | | 721 |
| Froot loops, kellogs | 128 | |
| Fullers earth | | 721 |
| Gasoline | 721 | |
| Gelatine, granulated | | |
| Gilsonite | 593 | |
| Glass bead | | |
| Glass cullett crushed | 1922 | |
| Gluten, wheat | | 561 |
| Glycerine | 1249 | |
| Golf tees | | |
| Graphite, ground | 400 | 481 |
| Grass seed | | 561 |
| Gravel | 1201 | 1362 |

| | | |
|---|---|---|
| Grits, corn | | 721 |
| Grits, rice | 673 | 721 |
| Gun powder | | |
| Gypsum, lump | 1442 | 1602 |
| Gypsum, powder | | 1281 |
| Hay | 80 | 384 |
| HDPE, polethylene | | 641 |
| Hominey | 593 | 801 |
| Hops | | |
| Hops, spent dry | 561 | |
| Hydrochloric acid | | |
| Ice, crushed | 881 | |
| Illmenite, ground | | |
| Iron chips | 2643 | |
| Iron ore | | |
| Iron oxide | 2883 | |
| Jet fuel, jp4 | | |
| Kafir | 641 | 721 |
| Kalsomine, powder | | |
| Kaoline, crushed | 320 | 352 |
| Kerosene | | |
| Lactose | 513 | |
| LDPE, polyethylene | | |
| Lead oxide | 481 | 2403 |
| Liginite | | 881 |
| Lima beans dry | 721 | |
| Lime, hydreated | | 481 |
| Lime, pebble | 881 | 1041 |
| Lime, quicklime | | 481 |
| Lime, slaked | 513 | |

| | | |
|---|---|---|
| Limestone, crushed | | 1522 |
| Limestone, dust | 1089 | |
| Linseed oil | | |
| Linseed, kernel | 400 | |
| Maize, kernel | | |
| Malt sugar | 481 | 561 |
| Malt, dry, whole | | 561 |
| Malt, ground, dry | 320 | |
| Malt, spent, damp | | 1041 |
| Malt, spent, dry | 160 | |
| Maltodextrin powder | | |
| Maple syrup | 1362 | |
| Marble, crushed | | 1522 |
| Menthol | 785 | |
| Metal dust | | 1922 |
| Methanol | 785 | |
| Methyl alcohol | | |
| Mica | 208 | 481 |
| Milk powder | | 320 |
| Milk sugar | 513 | |
| Miller, ground | | |
| Millet seed | 769 | |
| Mineral oil | | |
| Mineral spirits | 785 | |
| Molybdenum, floc | | 192 |
| Monosodium phosphate | 801 | |
| Muriate of potash | | |
| Mustard seed | 721 | |
| Naphthalene | | |
| Napthalene flakes | 721 | |

| | | |
|---|---|---|
| Navy beans, dry | | |
| Nitrate of soda | 1089 | |
| Nitric acid | | |
| Nitrocellulose | 400 | |
| Nylon | | 721 |
| Oat flour | 481 | 561 |
| Oat hulls | | 192 |
| Oat meal | 561 | 641 |
| Oat middlings | | 721 |
| Oats | 400 | 561 |
| Oats, bran | | |
| Oats, ground | 400 | 481 |
| Oats, rolled | | |
| Octane | 721 | |
| Oil, linseed | | |
| Oil, olive | 913 | |
| Oil, petroleum, crude | | |
| Oil, sperm whale | 913 | |
| Oil, transformer | | |
| Oil, turpentine | 865 | |
| Oxalic acid, crystals | | |
| Oyster shells, ground | 849 | |
| Paper, shreaded | | 192 |
| Paraffin wax | 721 | |
| PC, polycarbonate | | 577 |
| Peanut shell refuse | 64 | |
| Peanuts, shelled | | 721 |
| Peanuts, unshelled | 240 | 384 |
| Peas, dry | | 801 |
| Peat | 400 | 801 |

| | | |
|---|---|---|
| Perlite, expanded | | |
| Petroleum oil | 817 | |
| Phosphate rock, crushed | | 1281 |
| Phosphate sand | 1442 | 1602 |
| Plaster of Paris | | 881 |
| Plastic pellet | 545 | 769 |
| Ployethylene, pellet | | 577 |
| Ployvinyl chloride, powder | 481 | |
| Polyethylene pellet | | 593 |
| Polypropylene powder | 400 | |
| Polypropylene, pellet | | 577 |
| Polystyrene, expanded beads | 24 | |
| Polystyrene, pellet | | |
| Polyvinyl chloride, pellet | 769 | 833 |
| Popcorn, popped | | |
| Popcorn, shelled | 721 | 801 |
| Potash | | 961 |
| Potasium chloride | 32 | 48 |
| Potassium carbonate | | 801 |
| Potassium chloride | 1201 | |
| Potassium nitrate | | |
| Potassium sulphate | 673 | 769 |
| Potato flake | | |
| Potato starch | 641 | |
| Pumice | | 721 |
| PVC polyvinyl chloride | 769 | 833 |
| Quartz, sand | | 1602 |
| Rape seed | 721 | 801 |
| Rice | | 801 |
| Rice bran | 320 | |

| | | |
|---|---|---|
| Rice flour | | |
| Rice grits | 673 | 721 |
| Rubber, ground | | 801 |
| Rye | 705 | |
| Rye, flour | | |
| Salt, coarse crushed | 721 | 881 |
| Salt, granulated | | 1281 |
| Saltpeter | 1201 | |
| Sand, damp | | |
| Sand, dry | 1281 | 1602 |
| Sand, silica | | |
| Sandstone, crushed | 1281 | 1522 |
| Sawdust | | 192 |
| Sea water | 1025 | |
| Semolina | | 641 |
| Sesame seed | 432 | 593 |
| Shellac powder | | 561 |
| Silica flour | 561 | 641 |
| Silica gel | | 721 |
| Silica sand | 1522 | |
| Slag, furnace | | |
| Slakes lime | 513 | |
| Slate, crushed | | 1442 |
| Soap powder | 320 | 400 |
| Soda ash | | 721 |
| Sodium bicarbonate | 657 | |
| Sodium chloride | | |
| Sodium hydroxide, flake | 753 | |
| Sodium nitrate | | 1281 |
| Sodium sulphate | 1281 | |

293

| | | |
|---|---|---|
| Sorghum seed | | 801 |
| Soybean flour | 432 | 561 |
| Soybean hulls | | |
| Soybean meal | 577 | 801 |
| Soybean, flakes | | 400 |
| Soybean, whole | 753 | |
| Soybeean, cracked | | |
| Spelt flour | 400 | 481 |
| Starch powder | | 561 |
| Steel, chips | 2403 | |
| Sucrose - crystal | 1586 | |
| Sucrose - amorphous | 1507 | |
| Sugar, brown | 721 | |
| Sugar, dextrose, powder | | |
| Sugar, granulated | 849 | |
| Sugar, milk | | |
| Sugar, powdered | 801 | 961 |
| Sugar, raw | | 1041 |
| Sulfuric acid | 1794 | |
| Sulphur, crushed | | 1121 |
| Sunflower seed | 577 | |
| Talcum powder | | 993 |
| Tar | 1153 | |
| Tea leaves | | |
| Terephalic acid powder | 721 | |
| Timothy seed | | |
| Tin oxide | 1602 | |
| Titanium dioxide | | 801 |
| Tobacco, flake | 32 | 80 |
| Toulene | | |
| Transmission oil | 865 | |

| | | |
|---|---|---|
| Trisodium phosphate | | 961 |
| Urea, prill | 545 | 673 |
| Vermiculite ore | | |
| Vermiculite, expanded | 272 | |
| Walnut meats | | |
| Walnut shells, ground | 641 | 721 |
| Water | | |
| Wax | 240 | 320 |
| Wheat bran | | |
| Wheat gluten | 481 | 561 |
| Wheat, craked | | 721 |
| Wheat, flaked | 112 | 160 |
| Wheat, flour | | 561 |
| Wheat, ground | 641 | |
| Wheat, whole kernel | | 881 |
| Whey powder | 561 | 737 |
| Wood chips | | 481 |
| Wood flour | 240 | 400 |
| Wood shavings | | 160 |
| Xanthum gum | 769 | |
| Zinc ore | | |
| Zinc oxide | 160 | 481 |
| Zinc, calcined, crushed | | 1442 |

Slurry is a mixture of solids and liquid. The density of a slurry can be calculated as

$$\rho_m = 100 / (c_w / \rho_s + (100 - c_w) / \rho_l) \qquad (1)$$

where

$\rho_m$ = density of slurry (lb/ft³, kg/m³)
$c_w$ = concentration of solids by weight in the slurry (%)
$\rho_s$ = density of the solids (lb/ft³, kg/m³)
$\rho_l$ = density of liquid without solids (lb/ft³, kg/m³)

The slurry concentration by weight can be measured by evaporating a known weight of slurry and measuring the weight of dried solids.

295

## APPENDIX II:  Densities of Various Liquids

| Material | $(kg/m^3)$ |
|---|---|
| Acetone | 785 |
| Alcohol, methyl | 785 |
| Asphalt, liquid | 1041 |
| Ethyl ether | 705 |
| Gasoline | 721 |
| Glycerine | 1249 |
| Methanol | 785 |
| Mineral spirits | 785 |
| Octane | 721 |
| Oil, olive | 913 |
| Oil, sperm whale | 913 |
| Oil, turpentine | 865 |
| Petroleum oil | 817 |
| Sea water | 1025 |
| Sulfuric acid | 1794 |
| Transmission oil | 865 |

## APPENDIX III: Common Properties of Water, $H_2O$
### (www.engineeringtoolbox.com)

| $t$, C | Absolute pressure, P, $kN/m^2$ | Specific Heat, $Cp$, kJ/kgK | Specific enthalpy, H, kJ/kg | Prandtl's Number |
|---|---|---|---|---|
| 0.01 | 0.6 | 4.210 | 0 | 13.67 |
| 5 | 0.9 | 4.204 | 21.0 | |
| 10 | 1.2 | 4.193 | 41.9 | 9.47 |
| 15 | 1.7 | 4.186 | 62.9 | |
| 20 | 2.3 | 4.183 | 83.8 | 7.01 |
| 25 | 3.2 | 4.181 | 104.8 | |
| 30 | 4.3 | 4.179 | 125.7 | 5.43 |
| 35 | 5.6 | 4.178 | 146.7 | |
| 40 | 7.7 | 4.179 | 167.6 | 4.34 |
| 45 | 9.6 | 4.181 | 188.6 | |
| 50 | 12.5 | 4.182 | 209.6 | 3.56 |
| 55 | 15.7 | 4.183 | 230.5 | |
| 60 | 20.0 | 4.185 | 251.5 | 2.99 |
| 65 | 25.0 | 4.188 | 272.4 | |
| 70 | 31.3 | 4.191 | 293.4 | 2.56 |
| 75 | 38.6 | 4.194 | 314.3 | |
| 80 | 47.5 | 4.198 | 335.3 | 2.23 |
| 85 | 57.8 | 4.203 | 356.2 | |
| 90 | 70.0 | 4.208 | 377.2 | 1.96 |
| 95 | 84.5 | 4.213 | 398.1 | |
| 100 | 101.33 | 4.219 | 419.1 | 1.75 |
| 105 | 121 | 4.226 | 440.2 | |
| 110 | 143 | 4.233 | 461.3 | |
| 115 | 169 | 4.240 | 482.5 | |
| 120 | 199 | 4.248 | 503.7 | 1.45 |
| 125 | 228 | 4.26 | 524.3 | |
| 130 | 270 | 4.27 | 546.3 | |
| 135 | 313 | 4.28 | 567.7 | |

| $t$, C | Absolute pressure, P, kN/m$^2$ | Specific Heat, $Cp$, kJ/kgK | Specific enthalpy, H, kJ/kg | Prandtl's Number |
|---|---|---|---|---|
| 140 | 361 | 4.29 | 588.7 | 1.25 |
| 145 | 416 | 4.30 | 610.0 | |
| 150 | 477 | 4.32 | 631.8 | |
| 155 | 543 | 4.34 | 653.8 | |
| 160 | 618 | 4.35 | 674.5 | 1.09 |
| 165 | 701 | 4.36 | 697.3 | |
| 170 | 792 | 4.38 | 718.1 | |
| 175 | 890 | 4.39 | 739.8 | |
| 180 | 1000 | 4.42 | 763.1 | 0.98 |
| 185 | 1120 | 4.45 | 785.3 | |
| 190 | 1260 | 4.46 | 807.5 | |
| 195 | 1400 | | 829.9 | |
| 200 | 1550 | 4.51 | 851.7 | 0.92 |
| 220 | | 4.63 | | 0.88 |
| 225 | 2550 | 4.65 | 966.8 | |
| 240 | | 4.78 | | 0.87 |
| 250 | 3990 | 4.87 | 1087 | |
| 260 | | 4.98 | | 0.87 |
| 275 | 5950 | 5.20 | 1211 | |
| 300 | 8600 | 5.65 | 1345 | |
| 325 | 12130 | 6.86 | 1494 | |
| 350 | 16540 | 10.1 | 1672 | |
| 360 | 18680 | 14.6 | 1764 | |

## APPENDIX IV: Thermal Properties of Steam, $H_2O$
### (www.engineeringtoolbox.com)

| T, K | Density, g/cc | Heat Capacity, J/g.K | Thermal Conductivity, W/cm.K | Viscosity, g/cm.sec | Prandtl Number, Pr |
|------|------|------|------|------|------|
| 2000 | 1.09E-04 | 2.832 | 2.06E-04 | 6.27E-05 | 0.86 |
| 1500 | 1.46E-04 | 2.594 | 1.48E-04 | 4.91E-05 | 0.86 |
| 1000 | 2.20E-04 | 2.267 | 8.80E-05 | 3.43E-05 | 0.88 |
| 800 | 2.75E-04 | 2.130 | 6.60E-05 | 2.81E-05 | 0.91 |
| 600 | 3.66E-04 | 2.003 | 4.60E-05 | 2.14E-05 | 0.93 |
| 500 | 4.41E-04 | 1.947 | 3.65E-05 | 1.77E-05 | 0.94 |
| 400 | 5.55E-04 | 1.900 | 2.77E-05 | 1.40E-05 | 0.96 |
| 373.2 | 5.98E-04 | 2.020 | 2.48E-04 | 1.20E-04 | 0.98 |
| 370 | 5.38E-04 | 2.020 | 2.46E-04 | 1.19E-04 | 0.98 |
| 360 | 3.78E-04 | 1.980 | 2.37E-04 | 1.15E-04 | 0.96 |
| 350 | 2.60E-04 | 1.950 | 2.30E-04 | 1.11E-04 | 0.94 |
| 340 | 1.74E-04 | 1.930 | 2.23E-04 | 1.07E-04 | 0.93 |
| 330 | 1.14E-04 | 1.910 | 2.17E-04 | 1.03E-04 | 0.91 |
| 320 | 7.15E-05 | 1.890 | 2.10E-04 | 9.89E-05 | 0.89 |
| 310 | 4.36E-05 | 1.880 | 2.04E-04 | 9.49E-05 | 0.87 |
| 300 | 2.55E-05 | 1.870 | 1.98E-04 | 9.09E-05 | 0.86 |
| 290 | 1.42E-05 | 1.860 | 1.92E-04 | 8.69E-05 | 0.84 |
| 280 | 7.60E-06 | 1.850 | 1.86E-04 | 8.29E-05 | 0.82 |
| 273.2 | 4.80E-06 | 1.850 | 1.82E-04 | 7.94E-05 | 0.81 |

- T(K) = Temperature in Kelvin, t (C) = Temperature in Celsius
- t (C) = T (K) - 273
- Heat Capacity [=] 1 J/gK = 0.23864 BTU/lbm/F
- Thermal Conductivity [=] 1 W/cmK = 57.77 BTU/hr/ft/F
- Viscosity [=] 1 g/cm sec = 1 poise = 241.9 lbm/ft/hr

## APPENDIX V:  Common Properties of Air
### (www.engineeringtoolbox.com)

| Temperature, $t$, $^oC$ | Density $\rho$, $kg/m^3$ | Specific heat capacity, $Cp$, $kJ/kg\ K$ | Thermal conduct-ivity, $k$, $W/m\ K$ | Kinematic viscosity, $v$, $(m^2/s)\ x\ 10^{-6}$ | Prandtl's number, $Pr$ |
|---|---|---|---|---|---|
| -150 | 2.793 | 1.026 | 0.0116 | 3.08 | 0.76 |
| -100 | 1.980 | 1.009 | 0.0160 | 5.95 | 0.74 |
| -50 | 1.534 | 1.005 | 0.0204 | 9.55 | 0.725 |
| 0 | 1.293 | 1.005 | 0.0243 | 13.30 | 0.715 |
| 20 | 1.205 | 1.005 | 0.0257 | 15.11 | 0.713 |
| 40 | 1.127 | 1.005 | 0.0271 | 16.97 | 0.711 |
| 60 | 1.067 | 1.009 | 0.0285 | 18.90 | 0.709 |
| 80 | 1.000 | 1.009 | 0.0299 | 20.94 | 0.708 |
| 100 | 0.946 | 1.009 | 0.0314 | 23.06 | 0.703 |
| 120 | 0.898 | 1.013 | 0.0328 | 25.23 | 0.70 |
| 140 | 0.854 | 1.013 | 0.0343 | 27.55 | 0.695 |
| 160 | 0.815 | 1.017 | 0.0358 | 29.85 | 0.69 |
| 180 | 0.779 | 1.022 | 0.0372 | 32.29 | 0.69 |
| 200 | 0.746 | 1.026 | 0.0386 | 34.63 | 0.685 |
| 250 | 0.675 | 1.034 | 0.0421 | 41.17 | 0.68 |
| 300 | 0.616 | 1.047 | 0.0454 | 47.85 | 0.68 |
| 350 | 0.566 | 1.055 | 0.0485 | 55.05 | 0.68 |
| 400 | 0.524 | 1.068 | 0.0515 | 62.53 | 0.68 |

## APPENDIX VI: The Specific Heats of Some Common Liquids and Fluids (www.engineeringtoolbox.com)

| Product | (kJ/kg.K) |
|---|---|
| Acetic acid | 2.043 |
| Acetone | 2.15 |
| Alcohol, ethyl 32°F (ethanol) | 2.3 |
| Alcohol, ethyl 104°F (ethanol) | 2.72 |
| Alcohol, methyl. 40 - 50°F | 2.47 |
| Alcohol, methyl. 60 - 70°F | 2.51 |
| Alcohol, propyl | 2.37 |
| Ammonia, 32°F | 4.6 |
| Ammonia, 104°F | 4.86 |
| Ammonia, 176°F | 5.4 |
| Ammonia, 212°F | 6.2 |
| Ammonia, 238°F | 6.74 |
| Aniline | 2.18 |
| Benzene, 60°F | 1.8 |
| Benzene, 150°F | 1.92 |
| Benzine | 2.1 |
| Benzol | 1.8 |
| Bismuth, 800°F | 0.15 |
| Bismuth, 1000°F | 0.155 |
| Bismuth, 1400°F | 0.165 |
| Bromine | 0.47 |
| n-Butane, 32°F | 2.3 |
| Calcium Chloride | 3.06 |
| Carbon Disulfide | 0.992 |
| Carbon Tetrachloride | 0.866 |
| Castor Oil | 1.8 |
| Chloroform | 1.05 |
| Citron Oil | 1.84 |
| Decane | 2.21 |
| Diphenylamine | 1.93 |

| | |
|---|---|
| Dodecane | 2.21 |
| Dowtherm | 1.55 |
| Ether | 2.21 |
| Ethyl ether | 2.22 |
| Ethylene glycol | 2.36 |
| Freon R-12 saturated -40°F | 0.88 |
| Freon R-12 saturated 0°F | 0.91 |
| Freon R-12 saturated 120°F | 1.02 |
| Fuel Oil min. | 1.67 |
| Fuel Oil max. | 2.09 |
| Gasoline | 2.22 |
| Glycerine | 2.43 |
| Heptane | 2.24 |
| Hexane | 2.26 |
| Hydrochlor acid | 3.14 |
| Iodine | 2.15 |
| Kerosene | 2.01 |
| Linseed Oil | 1.84 |
| Light Oil, 60°F | 1.8 |
| Light Oil, 300°F | 2.3 |
| Mercury | 0.14 |
| Methyl alcohol | 2.51 |
| Milk | 3.93 |
| Naphthalene | 1.72 |
| Nitric acid | 1.72 |
| Nitro benzole | 1.52 |
| Octane | 2.15 |
| Oil, Castor | 1.97 |
| Oil, Olive | 1.97 |
| Oil, mineral | 1.67 |
| Oil, turpentine | 1.8 |
| Oil, vegetable | 1.67 |
| Olive oil | 1.97 |
| Paraffin | 2.13 |
| Perchlor ethylene | 0.905 |

| | |
|---|---|
| Petroleum | 2.13 |
| Petroleum ether | 1.76 |
| Phenol | 1.43 |
| Potassium hydrate | 3.68 |
| Propane, 32°F | 2.4 |
| Propylene | 2.85 |
| Propylene Glycol | 2.5 |
| Sesame oil | 1.63 |
| Sodium, 200°F | 1.38 |
| Sodium, 1000°F | 1.26 |
| Sodium chloride | 3.31 |
| Sodium hydrate | 3.93 |
| Soya bean oil | 1.97 |
| Sulfuric acid concentrated | 1.38 |
| Sulfuric acid | 1.34 |
| Toluene | 1.72 |
| Trichlor ethylene | 1.30 |
| Tuluol | 1.51 |
| Turpentine | 1.72 |
| Water, fresh | 4.19 |
| Water, sea 36°F | 3.93 |
| Xylene | 1.72 |

- $1 \ kJ/(kg \ K) = 0.2389 \ kcal/(kg \ ^{\circ}C) = 0.2389 \ Btu/(lb_m \ ^{\circ}F)$
- $T(^{\circ}C) = 5/9[T(^{\circ}F) - 32]$

## APPENDIX VII: The Specific Heats of Some Common Gases
### (www.engineeringtoolbox.com)

The specific heat at constant pressure and constant volume processes, and the ratio of specific heat and the individual gas constant - $R$ - for some common used "ideal gases", (approximate values at $68°F$ $(20°C)$ and $14.7$ psia $(1$ atm)):

| Gas or Vapour | Formula | $c_p$ (kJ/kg K) | $c_v$ (kJ/kg K) | $\gamma = c_p/c_v$ | $R = c_p - c_v$ (kJ/kg K) |
|---|---|---|---|---|---|
| Acetone | | 1.47 | 1.32 | 1.11 | 0.15 |
| Acetylene | $C_2H_2$ | 1.69 | 1.37 | 1.232 | 0.319 |
| Air | | 1.01 | 0.718 | 1.40 | 0.287 |
| Alcohol | $C_2H_5OH$ | 1.88 | 1.67 | 1.13 | 0.22 |
| Alcohol | $CH_3OH$ | 1.93 | 1.53 | 1.26 | 0.39 |
| Ammonia | $NH_3$ | 2.19 | 1.66 | 1.31 | 0.53 |
| Argon | $Ar$ | 0.520 | 0.312 | 1.667 | 0.208 |
| Benzene | $C_6H_6$ | 1.09 | 0.99 | 1.12 | 0.1 |
| Blast furnace gas | | 1.03 | 0.73 | 1.41 | 0.3 |
| Bromine | | 0.25 | 0.2 | 1.28 | 0.05 |
| Butatiene | | | | 1.12 | |
| Butane | $C_4H_{10}$ | 1.67 | 1.53 | 1.094 | 0.143 |
| Carbon dioxide | $CO_2$ | 0.844 | 0.655 | 1.289 | 0.189 |
| Carbon monoxide | $CO$ | 1.02 | 0.72 | 1.40 | 0.297 |
| Carbon disulphide | | 0.67 | 0.55 | 1.21 | 0.12 |
| Chlorine | $Cl_2$ | 0.48 | 0.36 | 1.34 | 0.12 |
| Chloroform | | 0.63 | 0.55 | 1.15 | 0.08 |
| Coal gas | | 2.14 | 1.59 | | |
| Combustion products | | 1 | | | |
| Ethane | $C_2H_6$ | 1.75 | 1.48 | 1.187 | 0.276 |
| Ether | | 2.01 | 1.95 | 1.03 | 0.06 |
| Ethylene | $C_2H_4$ | 1.53 | 1.23 | 1.240 | 0.296 |
| Freon 22 | | | | 1.18 | |
| Helium | $He$ | 5.19 | 3.12 | 1.667 | 2.08 |

| | | | | | |
|---|---|---|---|---|---|
| Hexane | | | | 1.06 | |
| Hydrochlor acid | | 0.795 | 0.567 | | |
| Hydrogen | $H_2$ | 14.32 | 10.16 | 1.405 | 4.12 |
| Hydrogen Chloride | $HCl$ | 0.8 | 0.57 | 1.41 | 0.23 |
| Hydrogen Sulfide | $H_2S$ | | | 1.32 | |
| Hydroxyl | $OH$ | 1.76 | 1.27 | 1.384 | 0.489 |
| Krypton | | 0.25 | 0.151 | | |
| Methane | $CH_4$ | 2.22 | 1.70 | 1.304 | 0.518 |
| Methyl Chloride | $CH_3Cl$ | | | 1.20 | |
| Natural Gas | | 2.34 | 1.85 | 1.27 | 0.5 |
| Neon | | 1.03 | 0.618 | 1.667 | 0.412 |
| Nitric Oxide | $NO$ | 0.995 | 0.718 | 1.386 | 0.277 |
| Nitrogen | $N_2$ | 1.04 | 0.743 | 1.400 | 0.297 |
| Nitrogen tetroxide | | 4.69 | 4.6 | 1.02 | 0.09 |
| Nitrous oxide | $N_2O$ | 0.88 | 0.69 | 1.27 | 0.18 |
| Oxygen | $O_2$ | 0.919 | 0.659 | 1.395 | 0.260 |
| Pentane | | | | 1.07 | |
| Propane | $C_3H_8$ | 1.67 | 1.48 | 1.127 | 0.189 |
| Propene (propylene) | $C_3H_6$ | 1.5 | 1.31 | 1.15 | 0.18 |
| Water Vapor Steam 1 psia. 120 – 600 °F | | 1.93 | 1.46 | 1.32 | 0.462 |
| Steam 14.7 psia. 220 – 600 °F | | 1.97 | 1.5 | 1.31 | 0.46 |
| Steam 150 psia. 360 – 600 °F | | 2.26 | 1.76 | 1.28 | 0.5 |
| Sulfur dioxide (Sulphur dioxide) | $SO_2$ | 0.64 | 0.51 | 1.29 | 0.13 |
| Xenon | | 0.16 | 0.097 | | |

- $\kappa = c_p / c_v$ - the specific heat capacity ratio
- $c_p$ = specific heat in a constant pressure process
- $c_v$ = specific heat in a constant volume process
- $R$- Individual Gas constant

305

## APPENDIX VIII:  Typical Values of Equilibrium Constants and Heats of Reaction  (Chemetron Corp., 1969)

$$CO + H_2O = CO_2 + H_2$$

| Temperature, C | $\Delta H_R$, J/mol | Kp |
|---|---|---|
| 93.3 | -40,868 | 4523 |
| 148.8 | -40,496 | 783.6 |
| 204.4 | -40,054 | 206.8 |
| 260.0 | -39,556 | 72.75 |
| 315.6 | -39,023 | 31.44 |
| 371.1 | -38,467 | 15.89 |
| 426.7 | -37,898 | 9.030 |
| 482.2 | -37,318 | 5.610 |
| 537.8 | -36,721 | 3.749 |
| 593.3 | -36,155 | 2.653 |

$$CH_4 + H_2O = CO + 3H_2$$

| Temperature, C | $\Delta H_R$, J/mol | Kp |
|---|---|---|
| 93.3 | 209,389 | $7.813 \times 10^{-19}$ |
| 148.8 | 211,729 | $6.839 \times 10^{-15}$ |
| 204.4 | 213,892 | $7.793 \times 10^{-12}$ |
| 260.0 | 215,862 | $2.173 \times 10^{-9}$ |
| 315.6 | 217,635 | $2.186 \times 10^{-7}$ |
| 371.1 | 219,230 | $1.024 \times 10^{-5}$ |
| 426.7 | 220,651 | $2.659 \times 10^{-4}$ |
| 482.2 | 221,910 | $4.338 \times 10^{-3}$ |
| 537.8 | 223,017 | $4.900 \times 10^{-2}$ |
| 593.3 | 223,954 | $4.098 \times 10^{-1}$ |

## APPENDIX IX: Specific Enthalpies of Elements in their Standard States, $H^o - H_f^o$ , kJ/kmol (Chemetron Corp., 1969)

| Temp, C | $H_2$ (g) | $O_2$ (g) | $N_2$ (g) | C (s, graphite) | S (g) |
|---|---|---|---|---|---|
| 0 | 7,755 | 7,932 | 7,948 | 849 | 8,160 |
| 37.8 | 8,844 | 9,041 | 9,048 | 1,165 | 9,383 |
| 93 | 10,458 | 10,679 | 10,669 | 1,721 | 11,225 |
| 149 | 12,079 | 12,358 | 12,295 | 2,375 | 13,112 |
| 204 | 13,703 | 14,061 | 13,928 | 3,115 | 15,033 |
| 260 | 15,328 | 15,794 | 15,575 | 3,936 | 16,982 |
| 316 | 16,954 | 17,554 | 17,236 | 4,827 | 18,957 |
| 371 | 18,587 | 19,348 | 18,917 | 5,780 | 20,948 |
| 427 | 20,222 | 21,169 | 20,615 | 6,787 | 22,951 |
| 482 | 21,864 | 23,013 | 22,337 | 7,841 | 24,970 |
| 538 | 23,509 | 24,881 | 24,079 | 8,934 | 26,998 |
| 649 | 26,819 | 28,684 | 27,621 | 11,.218 | 31,073 |
| 760 | 30,168 | 32,548 | 31,245 | 13,591 | 35,165 |
| 871 | 33,562 | 36,465 | 34,939 | 16,038 | 39,305 |
| 982 | 37,009 | 40,435 | 38,691 | 18,559 | 43,422 |
| 1093 | 40,505 | 44,443 | 42,494 | 21,153 | 47,562 |
| 1204 | 44,052 | 48,486 | 46,339 | 23,802 | 51.726 |

## APPENDIX X: Specific Enthalpies and Equilibrium Constants of Formation of Some Compounds above their Elements, $H^o - H_f^o + \Delta H_f^o$, kJ/kmol (Chemetron Corp., 1969)

| Temp, C | $H_2O$ (g) | | $CO_2$ (g) | |
|---|---|---|---|---|
| | $H^o - H_f^o + \Delta H_f^o$ | $Log_{10}K_f$ | $H^o - H_f^o + \Delta H_f^o$ | $Log_{10}K_f$ |
| 0 | -230,018 | 43.9218 | -384,974 | 75.4005 |
| 37.8 | ~228,753 | 38.3051 | -383,578 | 66.2567 |
| 93 | ~226,873 | 32.1369 | -381,415 | 56.2342 |
| 149 | -224.971 | 27.5823 | -379,131 | 48.8521 |
| 204 | -223,040 | 24.0794 | -376,740 | 43.1866 |
| 260 | -221,082 | 21.3002 | -374,.25] | 38.6999 |
| 316 | -219,091 | 19.0410 | -371,676 | 35.0585 |
| 371 | -217,060 | 17.1669 | ~369,025 | 32.0451 |

| 427 | -214,995 | 15.5878 | -366,301 | 29.5103 |
| 482 | -212,890 | 14.2380 | -363,514 | 27.3460 |
| 538 | -210,764 | 13.0713 | -360,670 | 25.4793 |
| 649 | -206,349 | 11.1545 | -354,831 | 22.4188. |
| 760 | -201,788 | 9.6453 | -348,816 | 20.0147 |
| 871 | -197,075 | 8.4261 | -342,662 | .18.0768 |
| 982 | -192,200 | 7.4197 | -336,382 | 16.4805 |
| 1093 | -187,190 | 6.5758 | -330,011 | 15.1434 |
| 1204 | -182,042 | 5.8574 | -323,547 | 14.0072 |

## Specific Enthalpies and Equilibrium Constants of Formation of CO (g) above its Elements, $H^o - H_f^o + \Delta H_f^o$, kJ/kmol. (Chemetron Corp., 1969)

| Temp, C | $H^o - H_f^o + \Delta H_f^o$ | $Log_{10}K_f$ |
|---|---|---|
| 0 | -105,940 | 25.8214 |
| 37.8 | ~104,840 | 23.2517 |
| 93 | ~103,.216 | 20.4419 |
| 149 | -101,586 | 18.3757 |
| 204 | -99,944 | 16.7917 |
| 260 | -98,.285 | 15.5379 |
| 316 | -96,608 | 14.5200 |
| 371 | 94,910 | 13.6770 |
| 427 | -93,187 | 12.9668 |
| 482 | -91,442 | 12.3590. |
| 538 | -89,677 | 11.8341 |
| 649 | -86,078 | 10.9707 |
| 760 | -82,403 | 10.2895 |
| 871 | -78,661 | 9.7373 |
| 982 | -74,860 | 9.2800 |
| 1093 | -71,011 | 8.8944 |
| 1204 | -67,119 | 8.5651 |

## APPENDIX XI: Enthalpies and Equilibrium Constants of Formation of Some Organic and Inorganic Compounds, $H^o - H^o_f + \Delta H^o_f$, kJ/kmol. (Chemetron Corp., 1969)

| Temp., C | Methane, $CH_4$ (gas). $H^o - H^o_f + \Delta H^o_f$ | $Log_{10} K_f$ | Propane, $C_3H_8$(gas) $H^o - H^o_f + \Delta H^o_f$ | $Log_{10} K_f$ | Ammonia, (gas) $H^o - H^o_f + \Delta H^o_f$ | $NH_3$ $Log_{10} K_f$ |
|---|---|---|---|---|---|---|
| 0 | -57,764 | 10.0893 | -68,605 | 5.77 | -29,557 | 3.560 |
| 37.8 | -56,438 | 8.3547 | -65,905 | 3.38 | -28,214 | 2.502 |
| 93 | -54,359 | 6.4122 | -61,360 | 0.64 | -26,161 | 1.317 |
| 149 | -52,107 | 4.9584 | -56,178 | -1.42 | -24,002 | 0.423 |
| 204 | -49,688 | 3.8206 | -50,388 | -3.05 | -21,762 | -0.279 |
| 260 | -47,081 | 2.9006 | -40,078 | -4.36 | -19,418 | -0.846 |
| 316 | -44,289 | 2.1393 | -37,251 | -5.45 | -16,975 | -1.314 |
| 371 | -41,319 | 1.4996 | -29,933 | -6.37 | -14,442 | -1.707 |
| 427 | -38.177 | 0.9543 | -22,185 | -7.16 | -11,818 | -2.042 |
| 482 | -34,869 | 0.4837 | -14,026 | -7.84 | -9,055 | -2.331 |
| 538 | -31,403 | 0.0726 | -5,499 | -8.43 | -6,238 | -2.584 |
| 649 | -24,030 | -0.6117 | +12,542 | -9.42 | -365 | -3.004 |
| 760 | -16,117 | -1.1581 | +31,736 | -10.20 | +5,841 | -3.337 |
| 871 | -7,790 | -1.6008 | +52,058 | -10.84 | +12,323 | -3.608 |
| 982 | +1,021 | -1.9693 | +73,199 | -11.36 | +19,066 | -3.832 |
| 1093 | +10.169 | -2.2799 | +95,059 | -11.80 | +26,093 | -4.021 |
| 1204 | +19,618 | -2.5438 | +117,535 | - | +33,299 | - |

| | | | | 12.18 | | 4.182 |
|---|---|---|---|---|---|---|

# INDEX

## A

Abstracts and Indexes .................... 23
activity ........................................ 128
activity coefficient ....................... 128
addition rule ............................... 121
algebraic methods ....................... 199
Anergy Method ............................. 41
Antoine's equation ....................... 130
arithmetic methods ...................... 198
*Articles of Association* ....................... 2
Assembly operations .................... 200

## B

*Basic research* ................................. 8
basis of calculation ...................... 197
Batch Process .............................. 202
Block Diagram Flowsheets ............. 91
*Break -Even Analysis* ....................... 60
Buyer's Guides and Trade Directories
.............................................. 23
By-Pass Process ........................... 205

## C

Capacity Ratio Exponent method .... 30
*Capital gains tax* ............................ 50
*Capital Investment* ......................... 19
capital ratio ................................. 26
*capitalised earning rate* ................. 63
chemical equilibrium .................... 125
chemical industry ............................ 3
chemical potential ........................ 127
Chemical Principle ........................... 6
Clausius - Clapeyron Equation ....... 155
closed system .............................. 201
Co-current Operation Process ....... 203
companies ...................................... 1
Composite Depreciation ................ 57
Concentration .............................. 115
conditions of thermodynamic
    equilibrium .............................. 125
Consecutive Operation Process ..... 203
Consulting .................................... 10

Continuous Process ...................... 202
Cost forecasting ............................ 24
Cost Indices .................................. 29
Counter - current Process ............. 204
Cox chart ..................................... 137
Customer Technical Service ........... 10

## D

Dalton's Law of Partial Pressures ........ 113
decision variables ......................... 218
Declining Balance Method ............. 52
Degrees of Superheat ................... 130
Density ........................................ 115
Depreciation ................................. 51
Dew Point ............................. 129, 142
differential heat of solution .......... 174
Direct Costs .................................. 19
*Discounted Cash Flow Rate of Return*
.............................................. 67
Dry Bulb Temperature .................. 142
Duehring plot ............................... 136

## E

Enthalpy ...................................... 124
Enthalpy concentration charts ...... 175
equilibrium constant .................... 176
equipment constraints .................. 220
Equipment Design ........................ 101
*Equipment Drawings* .................... 102
Equipment Flowsheet ..................... 91
Equipment Specifications ............. 103
Excise tax ..................................... 49

## F

flowsheet ...................................... 89
fugacity ....................................... 128

## G

*General and Administrative Expenses*
.............................................. 21
Gibbs-Duhem equation ................ 127

*grassroots project*............................ 11
Group Depreciation ........................ 57

# H

Haggenmacher equation .............. 156
Heat of Combustion ..................... 171
Heat of Formation ........................ 170
heat of fusion ............................... 154
Heat of Neutralisation ................... 172
Heat of Solution ........................... 172
Heat of Transition ......................... 153
heat of vaporisation ..................... 154
Heats of Ionisation and Dissociation
............................................... 175
Heats of Solvation and Hydration .. 175
Henry's law .................................. 138
Humid Heat ................................. 143
Humid Volume .............................. 143
humidity ....................................... 138
Humidity Charts ........................... 141

# I

Ideal Gas Volume .......................... 112
ideal solution ............................... 129
Improved Exponent Methods ......... 39
Income tax .................................... 50
Indirect Costs ............................... 19
inflation, ...................................... 80
information flow diagram ............. 253
Instrumentation and Control Diagram
............................................... 96
Insurance ..................................... 50
Inventory operations .................... 200
IUPAC convention ......................... 167

Key Products ................................... 7

# L

Lang factor .................................... 34
*Law of Amagat* ........................... 110
Law of Constant Heat Summation . 181
Law of Lavoisier and Laplace ......... 181
Lewis relation ............................... 145
*limited liability company* .................. 2
Literature Sources ......................... 13

# M

Manufacturing cost ........................ 44
*Manufacturing Expenses* ................ 20
Mass and Mole Ratios ..................... 109
Mass Fraction ............................... 107
Material and energy balances ....... 101
Material Balance Flowsheet ............ 92
mean heat capacity ...................... 123
mechanical equilibrium ................ 125
*Memorandum of Association* ........... 2
*Method of Constraints* ................. 217
modeling ...................................... 200
Molarity and Molality .................. 110
Mole Fraction .............................. 108
Mole Ratios ................................. 109

# N

net present value ........................... 66
*Non-Ideal Solutions* ..................... 134
Normal Boiling Point ..................... 129

# O

open system ................................. 201
Optimum profit ............................. 61
Order of magnitude estimates ........ 26

# P

Partial Pressure ............................. 112
payout time .................................... 64
*Performance and Design Parameters*
............................................... 198
phase rule ................................... 127
Physical Principle ............................ 5
Piping Flowsheet ............................ 93
Plant Engineering ............................ 9
Plant Operation .............................. 9
present value ................................. 66
Price/Cost Indices ........................... 23
Primary Raw materials ..................... 8
*private company* ............................ 2
process design ................................ 88
process design report ................... 279
*Process development* ....................... 9
Process engineering ......................... 9

process synthesis method............ 198
Product development ......................9
profitability of a proposed investment
...............................................59
Project Components ...................... 12
Project Engineering ..........................9
project risk analysis ........................ 82
Property taxes................................ 49
public company ...............................2
pure component volume ............... 110
Purge Process............................... 205
PVT systems................................. 126

R

Raoult's law................................. 130
Raw materials ..................................4
Recycle Process............................ 205
Reference Substance Plots............ 159
Return on Investment ..................... 62

S

Sales and Marketing ...................... 10
sales income ................................. 58
Saturation Temperature ............... 129
Scouting research .............................8
second law of thermodynamics .... 125
selling price.................................... 59
Separation operations .................. 200
simulation..................................... 200
Single Operation Process .............. 202
Sinking Fund Method...................... 55
Specific Heat Capacity .................. 118
split fraction coefficients .............. 254
Split Fraction Method ................... 252
standard heat of reaction ............. 169
Statistical Mechanics .................... 187
Study estimates.............................. 35
Sum of the Years Digit (SYD) Method

.............................................. 54
Superheated Vapour .................... 130
Surtax ........................................... 50
system boundary.......................... 197

T

tax.. ............................................. 49
Tax Depreciation ........................... 57
Technical Management ................. 10
TELOS........................................... 280
The Corresponding States Approach
.............................................. 189
The Energy Balance Flowsheet........ 93
The Perturbation Approach .......... 188
The Straight Line Method .............. 51
The Use of Equations of State ....... 189
The Use of Models........................ 188
thermal equilibrium...................... 125
Trouton's rule............................... 155
turn - key ....................................... 12

U

unit operations................................3
unit processes .................................3
Units of Production Depreciation .... 57
Units of Time Depreciation ............. 57
universal factor method ................. 27

V

Vapour Pressure........................... 129

W

Watson's Rule .............................. 158
wet bulb depression ..................... 145
Wet Bulb Temperature................. 144

www.ingramcontent.com/pod-product-compliance
Lightning Source LLC
Chambersburg PA
CBHW031806190326
41518CB00006B/213